THEY LAUGHED AT NOAH

Preparing For Natural Disasters

Kellye A. Junchaya

MEDCAP • Clifton, New Jersey

THEY LAUGHED AT NOAH
Preparing For Natural Disasters
By Kellye A. Junchaya

Published by: MedCap
                 P.O. Box 2085
                 Clifton, NJ 07015
                 medcap.pub@juno.com

Cover photo by Charles W. Banks. (Permission for use by Charles J. Banks) Waves from a Nor'easter crash over the seawall in Winthrop, Massachusetts, 1953.

ISBN 0-9666726-2-3

Cataloging-In-Publication Data:
Junchaya, Kellye A.
        They Laughed At Noah: Preparing For Natural Disasters/ by
        Kellye A. Junchaya.
                p. cm. illustrated
        Includes bibliographical information and index.
        ISBN 0-9666726-2-3 (pbk.)
        1. Natural Disasters 2. Earth Sciences 3. Weather I. Title
        GB5014.J
        904   LCCN: 98-92069

The earth expands, contracts and breaths,
And gives us beauty to feel and see —
But there is no justice, judge or jury
When Mother Nature sends her fury.

Dedicated to my wonderful husband Michael, who thought of the title, and to my three beautiful children, Megan, Madison, and Michael who are my inspiration.

Acknowledgments

There is no way to list inclusively all the people who contributed to this book and supported me, so I thank you all wholeheartedly. I am grateful to the many agencies, organizations, associations, academic institutions and groups that deal with natural disasters. In addition to the many authors listed in the bibliography, these groups made my research possible. I am especially indebted to The American Red Cross, Federal Emergency Management Agency and the International Association of Emergency Managers for their reviews and valuable input. The Church Of Jesus Christ Of Latter Day Saints has taught the principles of food and water storage and actively encouraged preparation since before I was born. I have been lucky to learn these lessons through the examples and teachings of my parents, Kendahl and Norma Johnson. They have been a huge support and I love and appreciate them. I also thank my publisher, Medcap and my very talented editor, Kendra Marsh.

Special thanks to the photographers and suppliers of the photographs used in this publication:

Charles W. Banks and Charles J. Banks (cover photo)
Andrea Booher (FEMA)
American Red Cross
United States Geological Survey
Norma Johnson

Contents

continued

Section 3 - Food And Water Storage

Introduction

The day before a big snow storm I went to the grocery store for a newspaper. I found myself jammed between hundreds of people frantically buying bread, milk, eggs, and other essentials. One lady told me that she normally drinks 1/2 gallon of milk in a week but was buying 3 gallons that day. According to friends and family, other grocery stores in the area were the same. That is when I realized three things. One, many people are unprepared for disaster or hard times. Two, people tend to hectically overcompensate when a situation arises that's severity, duration, and timing is uncertain. Three, there are not many books available to the public to teach them how to plan ahead and avoid the anxiety and frustration of last minute shopping and preparing.

Why is it human nature to wait until the last minute to buy extra food and get ready for emergencies? Is it because we don't think an emergency will ever arise that will affect us? I doubt it. Is it because we think we can always beat out the storm or other frantic shoppers? Probably not. Is it because we are so busy living our hectic lives that we can hardly plan for today, let alone tomorrow? Maybe. Is it because we do not have the space, money, time, energy or motivation to store extra supplies? Possibly. Is it because we just do not know how? I believe this is the main reason and it is the reason I decided to write this book. If people really knew how to be prepared and store extra food and water in small spaces and not have to spend a lot of money to protect their families and feel secure, more people would do it.

Preparing for natural disasters is like buying an insurance policy with one big advantage. When you buy automobile insurance, for example, you pay a certain amount of money each year to protect yourself financially in case you are involved in an accident. If you are lucky enough to make it through the year accident-free, the money is wasted. By wasted, I mean that it is gone and you never see it again. The money spent on premiums only comes back to you if you file a claim.

Spending time and money to protect yourself from nature's wrath is never wasted. If you never endure a natural disaster, your knowledge, supplies, and food storage are available to see you through the minor emergencies that happen to everyone. When a short term power failure occurs, you have flashlights, food that does not require cooking, and extra bedding for warmth. When you get a flat tire, you have a car kit that includes tools, flares, and snacks for the kids. When somebody in your family is injured, you have a well packaged and readily available first-aid kit. The list goes on and on because life is full of inconveniences. Preparing for the worst enables you to manage the minor troubles with ease.

All summer and fall, we watch squirrels collecting nuts and extra food for winter. The bees spend all their time making honey so they will have enough to feed themselves when food is scarce. Ants store food, spiders wrap up extra flies and bugs, camels store extra water for dry spells, beavers hoard food, and even plants keep extra stores of water and nutrients. It is time for people to start to prepare as well. You only need to look at a newspaper or listen to the news to hear of all the disasters in the world. Every day there are earthquakes, tornadoes, hurricanes, floods, fires, mud slides, erupting volcanoes, avalanches, tsunami, devastating storms, and much more. Even if disasters do not occur in your area, they can affect the economy and the food supply which is shipped to you.

The consequences of disasters alone should motivate you to start preparing now. Besides natural disasters, though, there are other emergencies such as chemical spills, evacuations, strikes, layoffs, broken automobiles, accidents, etc. With the proper planning and foresight, the bad times can be handled with relative ease. There are also every day inconveniences like plain old bad weather, sicknesses, unexpected company, bad traffic, extra busy days, overtime work hours, etc. If you are prepared, you will never have to go to the store on days like that.

The purpose of this guide is not to scare you into getting ready for disasters or for hard times that may arise in the future. The purpose is to motivate you to start planning ahead and to teach you how. I have also included detailed information about common natural disasters because understanding the mechanics of these events is pertinent. Knowing where, when, and how disasters happen and the damages they incur, helps you safeguard yourself and family. Understanding these events helps to control fear and shun dangerous actions. When you follow the guidelines given, the end result will be better protection for your family and peace of mind that you are ready to handle unexpected situations.

Section 1

Natural Disasters

Natural Disasters

"And every living substance was destroyed which was upon the face of the ground, both man, and cattle, and the creeping things, and the fowl of the heaven; and they were destroyed from the earth; and Noah only remained alive, and they that were with him in the ark." Genesis 7:23

Disaster is defined as a sudden or momentous event that causes distress, damage, and loss. The events that we normally call disasters (floods, earthquakes, tornadoes, hurricanes, etc.) would not be disasters if people, societies, and property were not in the way. A tornado in the middle of a barren, unpopulated desert would not cause distress, damage, or loss and therefore would not qualify as a disaster. An earthquake by itself is not a natural disaster. In fact, there are more than a million earthquakes each year around the globe — most of which we never feel. A tsunami that hits a deserted island is only a ripple in the ocean. Who cares if twelve feet of snow falls on some remote mountain? The events themselves are actually not the disasters. *The interaction of earth's natural happenings and human existence is, at times, disaster.* Our mere presence on the planet makes this violent interplay between man and nature inevitable. We cannot avoid, prevent, ignore, avert, or in any way escape natural disasters. We can diminish human loss and economic liability. We can prepare.

In a sense, we are living in denial of the certainties that are in our future. We build whole communities, sometimes more than once, on floodplains. We jam millions of people into the greatest seismic risk areas in the United States. We consider ocean front property prestigious rather than dangerous. Numerous mobile homes are parked in "Tornado Alley". Our first method of preparation for natural disasters should be gaining a healthy respect for our environment and recognizing our boundaries. The key to reducing loss of life and property is better planning and more preparation by individuals and their leaders.

As we change our personal, community, and government attitudes, we begin to reduce the misfortunate consequences of disasters. Traditionally, the individual attitude has been one of false security. "It won't happen to me." The community attitude has ignored the remote possibilities. "It won't happen here." The government has been reluctant to imagine the calamitous results of an event whose timing is uncertain. "I'll believe it when I see it." Another political attitude, especially when a request for funds is involved, has been to deal with the here and now. "I'll worry about that when it happens."

Some of these attitudes are beginning to change and better planning is the result. As we lose our tolerance for the bitter realities of unexpected events and lessen our acceptance of the risks, we start to make a real change. One way for us to continue these sound changes is to support or initiate legislation for improved regulations and higher building standards. When these regulations are passed, we need to bolster community efforts to enforce them.

Many government and independent groups are doing research into disaster prevention and preparedness. This research demonstrates the changing attitudes and helps to reduce losses when disaster strikes. For example, we have implemented many prediction techniques and complex warning systems which save countless lives. There are thousands of groups, sub-groups, associations, committees, organizations, agencies, and councils around the world working on disaster awareness, understanding, prevention, preparedness, education, and relief.

Judicious use of land, safer building standards, and early prediction and warning systems are vital to disaster preparedness. The value of these and other measures should not be underestimated. We should sustain the applause for the people who are making our environment safer through their hard work, lobbying, research, and experience. Unfortunately, it is not enough! There are still valid limitations that leave us at risk.

Since it is unreasonable, if not impossible, to restrict residences to low risk areas, zoning regulations are not a solution. Even if we could zone out the danger spots, there is no spot on earth that is without some risk of natural disasters. What do we do with the millions of people who already occupy land that is statistically unsafe? With the demand for land escalating, it is not prudent to ban areas in consideration of "what ifs".

Implementing improved building codes is difficult because it is often costlier to use safer construction methods. The higher costs are then passed on to the individual. (Although some of these costs may be recovered with lower insurance premiums.) Even if we could build every new structure with higher standards, they would not become "disasterproof". What about all the buildings and residences that were built before the adoption of regulations?

Predicting disasters and warning the public does not always have successful results. The possibility of danger obligates a warning to the public. How many times have possibilities for disasters turned into false alarms? Continually warning the public of potential hazards may have the same effect as the boy who cried wolf. When we are continually warned of disasters that never seem to hit, we begin to become numb and start ignoring the warnings.

Another difficulty is warning people during late night and early morning hours when the general population is asleep. It is difficult to warn large groups of people who are in theaters, malls, restaurants, and other public facilities. Even if these people received the warning, would they know where they could take shelter? We need to prepare ourselves and families. By having the necessary supplies, food storage, and family or neighborhood plans, we can lessen our losses, reduce anxiety, and assist the post-disaster relief efforts.

Avalanches and Landslides

"...every mountain and hill shall be made low..."
Isaiah 40:4

January 10, 1962, was a bright sunny day in the town of Ranrahirca, Peru, and the surrounding villages. The nice weather must have been enjoyed by the thousands of Peruvians living in the valley. The day before, several tons of snow fell on the already heavily covered mountains overlooking Ranrahirca, Yungay, and several villages. At 6:13 in the evening, some families may have been eating dinner, others were probably outside tending the animals, crops, and playing children.

Up on Mount Huascarán, South America's second highest peak at over 22,000 feet, the melting snow was breaking off the glacier. About three tons of snow and ice fell 3000 feet and roared down a canyon. In only two minutes, it completely wiped out two villages and claimed the lives of the inhabitants. Buildings, houses, livestock, and the bodies of 800 people now added to the falling snow and debris. Five other nearby villages also fell victim to the 50 foot high mass traveling at over 100 miles per hour. When the tons of snow and village rubble burst out of the canyon, it flattened and buried the town of Ranrahirca and its 2,700 inhabitants. In the seven minutes since the snow and ice broke loose from the mountain, over 3,500 lives were taken, eight towns and villages were wiped out, and a fertile valley of crops and animals was buried. The canyon walls dictated the direction of the snow and spared the town of Yungay.

Only eight years later on May 31, 1970, the Andes mountains again changed history. A World Cup soccer game, broadcast from Mexico City, was coming to a close. Doubtless many Peruvians were listening and rooting for their chosen favorite. At 3:23 on that Sunday afternoon, a massive earthquake shook the northern half of the country and turned many towns to rubble, including Yungay. Stunned and disoriented, the survivors emerged to inventory the damage. What they did not know was that ice and snow had been shaken loose from Mount Huascarán above them. The snow landed in mountain lakes and reservoirs which then overflowed and turned the side of the mountain into a flowing mass of mud, ice and rocks. They did not have much time to understand the noise that they must have heard. The 100 million cubic yard mass moved toward them at 180 miles per hour and completely covered the town of Yungay. The 20,000 inhabitants were not so lucky this time. They were buried alive. Only the tops of four palm trees that had marked the center of town were visible above the debris-ridden mud that was 60 feet deep in some places. The eight years of rebuilding Ranrahirca were in vain as it too was entombed along with other nearby villages.

Peru, South America

Although these terrible events in Peru did not allow time for evacuations or other life-saving actions, similar disasters are not destined for the same fate. In other words, mitigation in these instances may appear fruitless and may send the message that no matter what you do, destruction is inevitable at times. In fact, events like these demand attention. They are scrutinized for years and improvements in predictions, warning systems, evacuation procedures, construction methods, observation techniques, geological equipment, community disaster education efforts, and disaster technology are tangible and certain results. By continuing to heed warnings, educate ourselves, and prepare early, we can help to ensure that history does not repeat itself.

G. Plafker/USGS

The four palm trees from the center of Yungay are the only markers of the city and its 20,000 inhabitants, now completely buried below mud and debris.

G. Plafker/USGS
Aerial view of Mount Huascarán and the landslide that destroyed Yungay (left) and Ranrahirca (right) in 1970.

G. Plafker/USGS

The lower portion of the 1970 landslide that buried Yungay (front) and Ranrahirca (back).

G. Plafker/USGS

Block of Granodiorite that was transported by the 1970 landslide in Peru, and deposited west of Ranrahirca. It is estimated to weigh 7,000 metric tons! The pole at the base of the block is four meters high.

Avalanches

What They Are

Although the term avalanche applies to the fall of any material, it generally refers to snow and ice. The term landslide is used to describe the downhill movement of soil, rocks, mud, debris, etc. When a landslide or avalanche is caused by a volcanic eruption or is accompanied by lava, it is called a lahar.

There are many different types and forms of avalanches. An avalanche is categorized by the type of snow, how it breaks off, the quantity, the moisture content, how it flows, and several other factors. Classification of avalanches is difficult because the properties and behaviors of snow are not completely understood and avalanches are often mixtures of several types.

A general classification into two categories is made according to how the snow is dislocated from the slope and starts the avalanche. If the snow breaks off from one point and travels as individual crystals, it is considered to be *loose*. If the snow breaks off from a long horizontal fracture and travels as one body, it is called a *slab*. A loose snow avalanche, also called *pure* or *sluff*, normally starts from a small amount of snow (less than one cubic yard) and picks up more snow on the way down. The path looks like an inverted 'V' or a pear shape as the snow brings down adjacent crystals and fans out on the way down. This fanning out and picking up more snow is a process recently called "self organized criticality" by United States scientists.

A slab avalanche breaks away from a whole area at once and leaves a jagged wall of snow behind. The snow is cohesive and moves as one piece. Once the avalanche has started, however, it can break up into several slabs, chunks or even fine powder depending on the terrain and type of snow. The shape of the path is rectangular if the snow continues to slide as one body. This is generally the most dangerous type of avalanche.

Another criterion for classification is the depth of the snow that starts sliding. If only the top layer falls, it is known as a *surface* avalanche. (Most loose snow avalanches only involve the top layers.) If all the snow cover is involved right down to the bare ground, it is called a *full depth* avalanche. Full depth avalanches are common when the ground is smooth. Grass covered hills are typical underlying surfaces for full depth slides, especially if the grass is long.

The quality of the snow is either *wet* or *dry*. An interesting aspect of dry snow avalanches is the blast of compressed air that charges out in front flattening obstacles before burying them. This burst of air is seen on many videos where trees and barriers are knocked flat seconds before the snow even reaches them. The most common time for wet snow avalanches is in the spring when the warmth saturates the snow with melt water. When the snow is wet, it generally rolls into balls as it moves downhill and sets hard and heavy like concrete once it comes to a stop.

The slide may be down an open slope which is termed *unconfined*, or down a valley, canyon or gully, called *channeled*. Channeled avalanches usually pick up rocks, trees, shrubs, and other debris on the way down and can pick up speed and turbulence as the snow mass bounces off the retaining walls.

There are generally two ways in which the avalanche moves. It flows along the ground or it flies through the air. If the snow hugs the ground, it is usually simply termed a *flow* avalanche. If the snow is in the air, logically, it is called an *airborne* or *airborne powder* avalanche.

Because of all the conditions of an avalanche, the speed varies greatly. A dry, loose, surface avalanche can reach speeds of 225 miles per hour. A wet, full depth, slab would travel much slower at around ten to fifteen miles per hour.

How They Happen

There are many conditions that can contribute to the potential for sliding snow. If the wind blows the snow over the top of a mountain so it forms a cornice or overhanging "roof" of snow, even a little sun can start the ball rolling. Most avalanches occur soon after a heavy snow storm. The added weight on the surface overcomes the cohesive properties and the battle with gravity is lost. The properties of the snow and how well it binds to the lower layers and ground is also a significant consideration.

In determining the potential for avalanches, the steepness of the slope is an important detail. Most avalanches occur on slopes between 35 and 50 degrees. Slopes of over 60 degrees rarely accumulate enough snow for avalanches and slopes below 20 degrees have too slight an incline. The orientation of the slope is also meaningful as the amount of wind and sun are contributing factors. Other significant circumstances are the configuration and features of the terrain, the ground cover, and the altitude. (Most avalanches start between 6000 and 9000 feet.)

Once the environmental factors are optimal, an avalanche can be triggered by the vibrations from loud noises such as explosions, thunderclaps, gun shots, low flying aircraft, or even yelling. Earth tremors are also common avalanche instigators. Many unsuspecting people have set off an avalanche by walking or skiing on snow that is unstable. (Slab avalanches are made up of dense snow that may have a hardened surface and appear deceivingly secure.) There are also avalanche triggers that are more gradual, such as the breakdown of the snow structure, temperature rises over time, rain, and especially added snowfall.

The Destruction

Avalanches are deadly and can cover an entire town in a matter of seconds. They can have a mass of more than a million tons and move at speeds in excess of 200 miles per hour. Victims of this devastating disaster usually die if not found within the first hour. Statistics show that one in five victims are dead when the snow stops moving. After thirty minutes, a person has about a 50 percent chance of survival. In practice, only about 5 percent of avalanche victims are rescued alive. Unfortunately, there are approximately 150 people killed by avalanches each year in the world.

Death can occur quickly from the initial blow of snow and debris or injuries from the wrenching, buckling slide. Suffocation can cause death by inhalation of snow, lack of oxygen, or pressure on the chest which does not allow the lungs to expand. If victims are not rescued quickly, they will die from exposure to the cold and/or shock.

The Himalayan mountains have the greatest number of avalanches but not the highest death tolls. That distinction goes to the Alps. The destruction there is due to the large amount of avalanche paths (there are 10,000 in Switzerland alone) and the populated villages in the valleys below these mountains. The popularity of skiing has heightened tourism and expanded the building of hotels, ski lifts, roads, and residences in valleys. The potential for widespread damage is ever increasing.

Norma Johnson

One of the many villages nestled in the foothills of the Swiss Alps. The numerous avalanche paths and populated valleys there combine to rank the Alps highest in avalanche death tolls.

What We Can Do

There have been many efforts to protect communities from the damaging effects of avalanches. Some of these efforts have met with limited success but in general, the powers of nature overwhelm the resources of man. Forest cover is a very effective protection for valleys below. Unfortunately, many areas prone to avalanches have been cleared for lumber leaving the residents below at risk. Replanting of trees is difficult because young saplings are torn away by the sliding snow, but many reforestation projects are underway. Even so, a large avalanche (especially one that starts above the tree line) can flatten even a well-established forest.

Some valley towns have built barrier walls, snow-bridges, snow-rakes, or other support structures. These small fence-like structures are set at an angle to the mountain and are meant to hold back the snow cover. Although they may break up an avalanche that has already begun, they are really meant to prevent one from starting.

The various snow supports are somewhat effective for smaller avalanches but there are several criteria for success. They must be placed in the area where the avalanches start, at the proper angle and height and be spaced very closely to one another (both across and down the mountain). If the avalanche is large or starts at a higher altitude than where the barriers are placed, the snow charges over the barriers sometimes adding them to the debris. If the snowfall is deeper than the height of the supports, they are completely useless. The type of material is also carefully chosen to match weather conditions, expected pressures, ease of transport up the mountain, etc. These snow supports are very expensive to build, transport, install, and maintain. Their biggest advantage is the reforestation that is afforded by such measures.

Splitting wedges and channels have also been used to direct the snow away from specific buildings or areas. Wedges have been used for a long time but are somewhat outdated. If the snow is directed away from one building or structure, it tends to aim at another. In addition, it only provides safety for those inside the protected structure at the time of the avalanche. Channels are also limited because large avalanches tend to overflow them. Isolated buildings or roadways are often protected by galleries. These are simply lean-to roofs built overhead to allow the snow to slide past the structure without damaging it.

Another method of mitigation is to use explosives to set off controlled avalanches. While this practice is helpful, there is a tendency to use too much explosive. It is difficult to predict exactly how much snow will fall with gunfire or explosives and there have been times when the team of experts is astounded by the size of the intentional avalanche.

The objective is to set off small avalanches before enough snow can accumulate for a large destructive one.

Establishing and publicizing evacuation routes and educating valley residents about the dangers are essential precautions. In addition to the warning systems presently operating and continually improving, people must know what to do. Tourists (who may know very little about avalanches), valley residents, skiers, and other recreational visitors must be informed and prepared. There is not always time for evacuation since avalanches often occur with little or no warning. However, when conditions do forecast an impending avalanche or the movement of snow is slow enough, many lives can be saved with an efficient and speedy evacuation.

If you are hiking, skiing, climbing, camping, etc., in snowy mountain areas, read and obey all avalanche warning signs. Listen to weather reports and heed any advice regarding possible avalanches. If you are vacationing with your family, teach each member of your family about the hazards of avalanches and what to do. If possible, hire a guide who is familiar with avalanches and can help you avoid unnecessary risks.

It is a good idea to carry a radio transmitter when in avalanche territory so that you can be found if you become a victim. These are sold at sporting goods stores and specialty shops. Another good practice is to tie a brightly colored rope or cord around you to aid rescuers in finding you. The idea is to let the cord drag behind you as you go and if you are overtaken by an avalanche, hopefully some part of the rope will remain above the snow.

In freshly fallen snow, each person in your party should take a different path to avoid setting off an unstable mass of snow. In other words, do not follow each other's tracks. If you are downhill from an avalanche, take cover behind a tree or rock. Flatten yourself as much as possible and cover your head. If you are swept away, use your arms and legs in a swimming motion to stay on top of the snow. Try to keep your mouth shut tightly. In the event that you are buried, try to keep as much air space as possible around your face and chest. Try to save your oxygen supply by staying calm, refraining from struggling too much and only yelling for help if you hear rescuers nearby. If you see another person being swept away, watch him as long as you can so you can help rescuers locate him.

Successful rescues are contingent on swiftness. Searching a large area for buried victims can be slow, difficult and tedious. Rescue teams use different methods and detectors to try to locate victims as quickly as possible. Some of the detectors that might be used are sonar, radar, magnetometers (to detect skis), infrared detectors (body heat), detectors of the change in dielectric constant, and lasers. Rescue dogs are the most effective and can cover a large area surprisingly fast. A dog can search 1000 square feet in about a half an hour whereas a group of 20 rescuers would take about four hours to cover the same ground. Transmitters carried by victims are also very effective provided rescue teams with receivers are nearby.

Landslides

What They Are

Falls, slides, and flows of the ground under the force of gravity are described by the term *landslide*. Like avalanches, landslides are categorized by their characteristics. There are many different types of landslides so they are not easily predicted or prevented. There are general descriptions of different types of landslides and there are warning signs and likelihoods but there are no definites. In other words, seemingly stable ground can still be at risk and unstable ground can stay put for years.

When all the layers of soil on a hillside gradually slip, it is known as *slumping*. As the hillside gradually slides down, the bottom of the hill becomes level or slightly tipped upward. Slumping is evident in trees, poles, fences, walls, and other vertical structures that start to tilt or move. Slumping can occur fairly rapidly or take place over many years.

If only the top few inches of soil gradually move downward with the pull of gravity, it is *creeping*. This usually occurs on mountains or hills that have a fairly steep slope. After heavy rains or floods when the soil is supersaturated and starts to slide, it is a *mudflow*. Hillsides that experience creeping are more prone to mudslides after rains or landslides after a trigger (such as an earthquake).

The pumping of water from underground can cause the ground above to collapse or fall vertically. Mining, construction, quarrying, and other forms of underground excavations deplete the ground of its subterraneous support. Seismic action can also cause the loss of underground support. The result is a perpendicular decent of earth known as *subsidence*.

When subsidence occurs suddenly and rapidly, it is commonly referred to as a *sinkhole*. Sinkholes occur when groundwater dissolves supporting substrata such as limestone or gypsum. In 1993, a sinkhole formed under a parking lot of a hotel in Atlanta, Georgia. The 100-foot wide, 25-foot deep hole swallowed up several cars and killed two people.

The speed of a landslide varies greatly. Factors affecting the rate of flow are the steepness of the slope, the amount of water in the soil, whether it is flowing openly or down a canyon, ravine or channel, the substance that is falling (mud, dirt, rocks, lava, debris, mixture, etc.), and any barriers that impede progress (natural and man-made). A landslide can be as slow as one centimeter a year or it can reach speeds in excess of 200 miles per hour.

Andrea Booher/FEMA

The 1998 rains associated with El Niño caused landslides that destroyed millions of dollars worth of homes in California including these in Pacifica..

How They Happen

Any force that increases the weight of the topsoil, lubricates the layers of substrata, or makes the slope too steep can cause a landslide. Heavy rainfall or snow melts can both add weight to the soil and lubricate underlying layers. Rapid temperature changes also start the land moving by expanding and shrinking soil formations and creating ice heaves between layers. The wave action of oceans often changes the slope of nearby mountains and hills allowing gravity to take over.

Man also has to share the blame. The deforestation of hillsides greatly increases the potential for landslides. The roots of trees trap water and help anchor the soil in place. Without trees, the soil is free to start a downhill movement with only a nudge from nature, gravity, or man. Forest fires also deny the slopes their trees, shrubs, and other stabilizing vegetation. Mining, quarrying, and road construction diminish the underlying support and solidity of the land.

Some of the most devastating landslides have been initiated by other natural disasters. Earthquakes, volcanic eruptions, tornadoes, hurricanes, and storms can all provoke landslides. A little more subtle and indirect is global warming which leads to weather changes that can also contribute to landslides.

The Destruction

A landslide disaster occurs somewhere in the world almost every day! Every state in America is at risk from landslides. The annual death rate from this catastrophe in the United States is between 25 and 50. The annual loss from landslides in the U.S. is estimated to be between one and two billion dollars. A mammoth landslide near a body of water can incite a tsunami. Landslides can also produce flooding by filling rivers and lakes with debris. The water then bursts its banks and drenches towns and villages or generates more landslides.

In 1983, heavy rains caused a huge mudslide which damned Spanish Fork Canyon in Utah. An immense lake was created which flooded the town of Thistle. The mud mass was about a half-mile deep and in some areas one mile wide. It was the most expensive landsliding event in United States history costing around 500 million dollars.

Utah

Spanish Fork Canyon

Although landslides can be slow and gradual, they can also be sudden and catastrophic. The latter kind usually occurs with no warning. A 1985 report on landslides by the Committee on Ground Failure Hazards of the National Research Council states, "The loss of life from landsliding is comparable to the total loss of life from floods, earthquakes and hurricanes."

What We Can Do

One of the best ways to reduce losses from landsliding is to have good land management. As valleys become crowded and polluted, people start moving up the mountains. It is natural to want a little space, privacy, fresh air, and a beautiful view of the city below. However, we must utilize caution and employ prudent judgment when building permits are requested for mountainside locations.

Residences in valley regions or on hillsides require special precautions. Get a ground assessment for your property. A geotechnical expert can give you personalized advice for your home and property. The county geologist or the planning and development department usually have information about areas that are prone to landsliding. Since many landslides occur over and over in the same places, it is a good idea to contact the county geologist before buying property on or below a hillside.

Supposing you already live in an area that has a potential for landslides, there are a few things you can do ahead of time. Plant trees, grass and other types of ground covering vegetation on the unstable surfaces. Put up barriers or deflecting walls to divert the mud flows away from property and play areas. Be careful not to direct the flow onto another person's property or you may be liable for any damages caused by the landslide. It is best to work with your neighbors and community as a group to avoid disastrous results. Landslides and mudflows are covered by the National Flood Insurance Program so make sure you have the proper insurance. Talk to your insurance agent to make sure all potential damages would be covered by your policy.

Other mitigating steps can be taken by your community government or land management agencies. Trenches are sometimes effective in "catching" and stopping smaller landslides. Do not try to make your own, as digging in unstable ground can trigger a slide. Underground drainage helps to keep the water from weighing down the side of slopes. Government should also control quarrying and mining so the underground support is not lost. A warning system and evacuation plan should be implemented in any towns or areas where landslides could be a problem.

When you are enjoying recreational activities in the mountains or valleys, follow the same advice given previously for avalanches. Try to walk on high ground and stay away from gullies and channels. Climb to higher ground as soon as any landsliding begins. If escape is impossible, roll into a ball and cover your head.

If you are indoors, stay inside. Take cover under a sturdy desk or table. Make sure you have your own evacuation routes (more than one) planned out. Have a family plan, evacuation kits, car kits, and emergency supplies (see Section 2). Be prepared with an adequate food and water supply as travel and communication lines could be down for some time following landslide damage. Also, food shipment and water purity may be affected by the landslide or flooding that landslides cause (see Section 3).

Cyclones, Typhoons and Hurricanes

"...and a great whirlwind shall be raised up from the coasts of the earth." Jeremiah 25:32

The deadliest natural disaster in American history was a hurricane. In 1900, the city of Galveston, Texas was a booming resort area and claimed one of the busiest ports in the United States. Galveston boasted many attractive qualities that caused over 38,000 people to call this area home. The warm climate, the protected harbor, the beautiful beaches, and the expanding business opportunities blinded the citizens to its reality. The city was built on a one mile wide sandbar whose highest point was less than nine feet above sea level. The day and night of September 8, 1900, opened the eyes of those who survived.

The day began with huge waves pounding the shore and flooding the nearest streets. Many spectators were on the beaches watching the thrilling sight. Isaac Cline, chief of the United States Weather Bureau office in Galveston, warned the sightseers to get to higher ground. There was no fancy warning system at that time. He rode his horse to the beach calling out to the unsuspecting wave watchers.

The winds became heavier as the morning hours passed and the rain began to fall. By noon, the winds were at 30 miles per hour, the rain was heavier, and the barometer was dropping. Practically everyone knew a storm was brewing but it was a little late to start preparing. By mid-afternoon, half of the city was under water and people were clambering for shelter and high ground. The bridges to the mainland were already washed out so the residents and tourists had no escape. By 4:00 P.M., the winds had reached hurricane force, the telegraph and telephone lines were down, and all of Galveston was submerged. At the highest point, people were wading in a foot of water while in other spots people were struggling neck deep. By evening, the estimated wind speed was 120 miles per hour or more. (The anemometer had blown away earlier when the wind was sustained at 84 miles per hour with gusts up to 100 miles per hour.) Many homes in Galveston had heavy slate shingles which became deadly missiles when the wind dislodged them. The water rose another nine feet between 7:30 and 8:30 P.M. The storm continued battering the town and killing its residents and tourists until midnight. The tide scoured and ravaged the city once more on the retreat.

The nightmare was only beginning. A stifling heat followed the storm and added to the gloom and disbelief of the survivors. Isaac Cline wrote in his official report to the U.S. Weather Bureau, "Sunday, September 9, 1900, revealed one of the most horrible sights that ever a civilized people looked upon." Galveston was completely destroyed, costing an estimated 30 million dollars in damages. More people lost their lives than in any other natural disaster in the United States, before or since. More than 6000 people perished in Galveston that night. At least 1000 more people died in other Texas cities along the Gulf of Mexico. The storm left 8000 people homeless. Famine and water shortages threatened the demolished city. Relief did not arrive for five days because of the destroyed bridges. The overwhelming number of corpses caused a burial problem. Large barges took the dead out to the Gulf where they were weighted down and buried at sea. The bodies soon started washing up on the beaches, adding insult to injury. It took weeks to then cremate the unfortunate victims of the tragedy.

Drawn by F. Cresson Schell for "Leslie's Weekly" / American Red Cross
1. The splendid city of Galveston in the wild fury of the hurricane. 2. Wrecks of shipping and grain elevator at the Galveston wharf. 3. The Galveston Strand, a principal street, in the height of the storm. 4. Galveston cut off from railroad communication by wind and wave.

Drawn by G.W. Peters from Telegraphic Reports/American Red Cross

"The Destruction of Galveston"

A year later, the wiser citizens started constructing a seawall to hold back the storm tides of the future. The protective wall is seventeen feet above mean low tide, sixteen feet wide, almost eight miles long and cost nine million dollars to build. They also raised and filled the town to reach a higher altitude. The southern gulf side was elevated to 17 feet to be even with the seawall. The land slopes gently down to sea level on the northern side. The next hurricane to hit Galveston was on August 16, 1915.

Although the estimated property damage was 50 million dollars (mostly to the causeway), only eight people died and the city survived.

Texas

Galveston

The costliest natural disaster in United States history was also a hurricane. Hurricane Andrew hit Florida in 1992 and managed to tally up about 30 billion dollars in damages! The future costs of natural disasters will certainly continue to climb. The death toll was only 58 (18 of which occurred during relief efforts) but as much as a million people had to evacuate their homes.

Those who were not evacuated or returned early, had several days (sometimes weeks) of pure misery. Besides the obvious despair of losing loved ones and property, other maladies tortured the already ravaged region. The temperatures following Andrew were in the nineties during the day and eighties at night with nearly 100 percent humidity. With no electricity in many towns, there were no air conditioners, fans, or coolers. With little or no trees, there was no shade. With screens and screen doors long blown away, opening doors and windows was unthinkable. The mosquitoes and other insects of southern Florida's tropical climate would make life unbearable. Water supplies were contaminated and there was no food available anywhere in the area. Those who had gas grills but had not filled their tanks were out of luck. The electricity outage made it impossible for propane storehouses to pump the gas from the tanks. As is often the case following a natural disaster, many survivors related their shock and surprise at the severity of the storm. People who have lived through a natural disaster often say that the terror and strength of nature's wrath goes beyond what they had imagined. Even with all the early warnings and previous history of hurricanes, people were amazed at the strength and destruction of hurricane Andrew. (Fifty-seven hurricanes hit the state of Florida between 1871 and 1989 — the most of any state.)

Florida

Homestead

On November 13, 1970, a cyclone in the Bay of Bengal hit Bangladesh (then known as East Pakistan) and caused massive destruction. It was one of the deadliest natural disasters in the world. It wiped out over a million cultivated rice paddies, destroying an estimated 800,000 tons of rice. It drowned hundreds of thousands of animals and livestock. The combination of massive flooding, high winds, poorly built houses, little warning, no evacuation routes or transportation, delayed and unenthusiastic relief efforts, contaminated water, and famine caused a million people to die.

Bangladesh

The deadliest natural disaster in America, the costliest natural disaster in America, and one of the deadliest natural disaster in the world were all hurricanes. In my opinion, this qualifies hurricanes as the most destructive of all natural disasters.

What They Are

A tropical storm with wind speeds at or above 74 miles per hour is called a cyclone, hurricane, or typhoon, depending on its location. In the Atlantic Ocean and east of the international date line in the Pacific Ocean, it is called a hurricane. The name comes from a Taino Indian word meaning "big wind". In the Pacific Ocean, west of the international date line, it's called a typhoon. This name comes from a Chinese word meaning "great wind". In the Indian Ocean, a tropical storm is called a cyclone which comes from a Greek word meaning "coil". The word *cyclone* is also used to define the general phenomena of a large circular atmospheric system. By this definition, swirling storms in the arctic and tornadoes as well as tropical storms are all types of cyclones. In this chapter, we are discussing just the *tropical* storms known as hurricanes, typhoons, and cyclones. In structure, strength, wind speed, damage, moisture content, development, time of year they occur, frequency, and any other attribute, they are identical. Only the region where they occur causes the name to change.

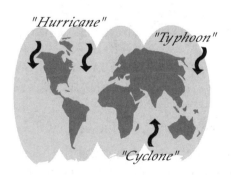

How They Are Formed

Since the sun heats the earth's surface unevenly, the atmosphere develops areas of low and high pressure. The warm, lighter air rises (low pressure) and the heavier cold air sinks (high pressure). When a low pressure area develops over warm tropical oceans, the rising air contains a lot of moisture. The earth's rotation affects the air and the Coriolis force keeps it from rising straight up. Instead, the humid air spins as it rises — counter clockwise north of the equator and clockwise south of the equator. The air cools as it rises and the moisture in the air condenses. This condensation produces rains and releases heat energy which fuels the system. (The energy a hurricane produces in one day would be enough to supply the entire country with heat and light for a year!)

After the air rises, the void is filled with the surrounding warm air which then rises as well. As the outside air rushes in to fill the space left by the rising air, the wind speeds increase. This phenomenon is governed by the conservation of angular momentum. That is the same law that causes an ice skater to spin faster as she brings her outstretched arms in to her chest. As long as the system continues to draw in warm moist air that rises and condenses, the storm continues to brew.

A center of calm, low pressure — the "eye" — is surrounded by ferocious winds. The eye of the storm can be anywhere from 14 to 100 miles wide with the average between 20 and 30 miles wide. The winds can reach diameters of 300 to 400 miles with 100 miles the average. The greater the difference in pressures inside and outside the storm, the greater the wind speeds will be.

A tropical disturbance has winds of less than 25 miles per hour; a tropical depression has wind speeds of 26 to 38 miles per hour; and a tropical storm has winds of 39 to 73 miles per hour. Once the wind speeds reach 74 miles per hour the storm is classified as a hurricane. There are five categories of hurricanes depending on the wind strength. A category five hurricane has wind velocities of over 200 miles per hour and the destruction is catastrophic. Hurricane Andrew was a category five.

When the hurricane hits land or cold water, the air that rushes in to fill the space of the rising air, is colder, heavier, and contains very little moisture. Since this air does not condense, the fuel or energy for the storm is lost and the hurricane slowly dies out.

How They Move

There are many complicated factors and inter-related atmospheric conditions that produce and drive hurricanes. That is why they are unpredictable, and finicky. Their path is determined by ocean currents, wind currents, atmospheric conditions (hurricanes avoid areas of high pressure), and by their own energy. Computer models try to predict the direction hurricanes will take based on information from weather stations, satellites, surveillance flights, and local observations. With so many factors to consider, even computers have to simplify or generalize some of the earth's conditions and the result becomes an educated guess.

We do know that hurricanes form between five degrees and thirty degrees north and south of the equator. The storm moves an average of 200 miles a day and can follow a path of 3000 miles. The average life span of a hurricane is nine days. The surface temperature of the ocean needs to be at least 77 degrees Fahrenheit (25 degrees Celsius) and warm moist air is a requisite. All the necessary conditions are present and ready to produce tropical storms between June 1st and November 30th — the official hurricane season.

Naming Them

The hurricanes and tropical storms of the early years were named for distinguishing characteristics. A storm may have been named for the holiday it occurred on, the city it destroyed, the year it happened, or other identifying attributes. During World War II, hurricanes and tropical storms were named using the phonetic alphabet (Able, Baker, Charlie...). As this was confusing, the practice of officially christening storms and hurricanes with female names started in America in 1953. The names were given in alphabetical order with each successive storm. (Q, U, X, Y and Z are not used for naming since there are not many names beginning with these letters.)

Complaints poured into the National Weather Bureau by both men and women. The women complained that having such a destructive and feared force named after them was an insult. The men complained that one of nature's biggest and strongest phenomena should be named after a man. So in 1978, the eastern Pacific storms were named both masculine and feminine. The next year the hurricanes and tropical storms of the Atlantic followed suit and included men's names.

The World Meteorological Organization has six bisexual lists of names that are taken from English, Spanish and French. Every six years the lists repeat themselves with a few exceptions. If a storm is especially destructive or unique, the name is retired and only used for that particular storm forever after. There are also different sets of names for different parts of the world.

The Destruction

The devastation caused by a hurricane is unimaginable. The low atmospheric pressure and strong winds combine to cause a huge storm surge. This surge, or wall of ocean water that charges the land, can be as high as 30 feet and can contain millions of tons of water. Ninety percent of all hurricane deaths are caused by the storm surge and its accompanying floods. The waves that are escorted in by the hurricane inevitably flood cities, damage property, and erode beaches. One day of hurricane wave action has the equivalent effect on the beaches as 200 years of normal waves. The characteristic torrential rains overflow rivers and creeks and bring on more flooding which can then incite landslides. It is ironic that with all the flooding, the water supply for drinking usually becomes contaminated and a water shortage ensues.

The powerful winds of a hurricane uproot trees, demolish houses and turn debris into deadly missiles. The winds also feed fires that are started by broken gas and electric lines. Fire stations may not even hear of the blazes since communication lines are usually damaged. Even if fire trucks can get to the fires through the damaged city, the wind often blows the water from hoses away from the fire. Hurricanes also spawn tornadoes that blast new trails of destruction. The energy released from a hurricane each day is equal to 500,000 atomic bombs!

Andrea Booher/FEMA
Hurricane Marilyn showed how destructive hurricanes can be when it hit St. Thomas in 1995. (Marylin was a category 3.)

What We Can Do

More than 70 million Americans are at risk from Hurricanes. Since coastal cities are becoming more and more populated, efficient and effective evacuation plans are essential. The city or county should have a warning system, evacuation plan, and relief program. If you live within 100 miles of the coast, contact the emergency management office or American Red Cross chapter in your area to find out about the regional warning and evacuation plans. You also need to have your own family plans, emergency supplies and evacuation kits (see Section 2).

Before hurricane season starts, install storm shutters or cut plywood (1/2 inch marine plywood is best) to fit your windows. Label the plywood with the corresponding window and drill holes for screws every 18 inches or so. Store them in an easily accessible place or get them out at the beginning of hurricane season. You may also consider having hurricane straps, which hold the roof to the walls, installed by a professional. They are fairly inexpensive and greatly reduce the chances of your roof flying off your home in high winds.

Get good insurance and provide your insurance agent with a list of your belongings and their worth. Homeowners insurance does not cover flooding from a hurricane. Ask your agent about the National Flood Insurance Program and make sure you are covered for any kind of damage that a hurricane could inflict. It is a good idea to make a video or take pictures of your house, property, and possessions. Save receipts for all big purchases and keep them with your important papers.

Keep your trees trimmed of all dead branches. Make a plan for your pets in the event of an evacuation. Take first-aid classes and attend community council meetings where hurricane preparedness is discussed.

At the beginning of hurricane season, obtain all your supplies and get them organized. It is almost impossible to get basic supplies when a hurricane is imminent. Even if you can find what you need, you will certainly have to stand in line and fight crowds of other unprepared citizens. A large amount of anxiety can be relieved by simply having everything you need ahead of time. When a hurricane watch or warning is issued, you can spend your time securing your property, gathering your family, and getting ready for a possible evacuation. Do not give yourself the unnecessary headache of having to run around for supplies at the last minute!

Fill your propane tanks. If you use your outdoor grill frequently during the summer, you may need to fill the gas tank more than once. Watch, listen to, or read the news frequently so you will be aware of any hurricane watches or warnings. You can purchase a NOAA (National Oceanic and Atmospheric Administration) weather radio for about $40 to get the latest updates. It is equipped with an alarm to alert you to watches or warnings.

When a hurricane watch is issued, it means that hurricane conditions are possible within 24 to 36 hours. This is the time to start securing your house. Board up your windows and remove antennas and turbine roof fans. (Cover the hole that the detached fan exposes.) Bring in or anchor outdoor furniture, toys, tools, etc. Double check your supplies and food storage. Turn your refrigerator and freezer to their coldest settings and open them as little as possible. Food will then stay cold for a day or two if there is an electricity outage. Fill pots, jugs, bathtubs, and other containers with water in case your water supply becomes contaminated. If you own a boat, secure it to the moor or trailer. If it is on a trailer, use tie downs to secure your trailer to the ground or house. Keep your family together if possible. Make sure you have fuel in your car and start getting ready to evacuate.

When a hurricane warning is issued, it means that hurricane conditions are expected in less than 24 hours. At this point, your home should be secure and ready. If you live in an area prone to flooding, raise furniture or transfer it to a higher floor of the house. Listen to hurricane reports and follow all instructions. Remember, you do not need to wait for an official or forced evacuation to leave. The sooner you leave the area, the less traffic and confusion you will encounter. If you live in a mobile home, you would be wise to evacuate the area as soon as a warning is issued.

Turn off electricity and the main water valve. Lock up and take your pre-made evacuation kits to a shelter. When an official evacuation is ordered, leave immediately. Many people feel that they need to stay with their homes and belongings and are very reluctant to evacuate. This is a very dangerous attitude! If you want to "protect" your things, forget it. Your presence will not change the effects of the forces of a hurricane. Staying to enjoy the wild ride of nature's energy could cost you your life. There are many reports of disastrous hurricane parties and needless loss of lives.

In the event that you are not evacuated and are at home when the hurricane hits, stay inside. The safest place to be is a basement or storm cellar if you have one. If not, stay in the center of the house away from windows and doors. If you live near the shore, flooding may preclude staying in a basement or storm cellar. In case you are at work, school, or in another public building, follow procedures and move to the center of the building.

Do not use candles or open flames for lighting if the electricity goes out. When the eye of the hurricane passes, do not make the mistake of thinking the storm has ended. The calm could last fifteen minutes or one to two hours but the other half of the fury is on its way. By being prepared with knowledge, emergency supplies, extra food and water, and a well-secured house, it should be easier to stay calm. Check for damages following the hurricane or when you return to your home. If you smell gas, open the windows and go outside right away. Shut off the gas from the outside main valve and call the gas company from a pay phone or neighbor's house. Once you turn off the gas, you must have a qualified or trained person turn it back on for you. If you suspect electrical damage (sparks, smell of burning insulation, frayed or broken wires, etc.) turn off the electricity at the circuit breaker or fuse box. Avoid and report broken or dangling power lines outside. Be careful of poisonous snakes and insects that seek higher ground after flooding. Try not to drive as there are flooded roads, washed out bridges, and rescue workers trying to get through. Start making repairs and getting ready for the next hurricane!

Norma Johnson
Great Sand Dunes National Monument, Colorado.

Drought and Famine

"And then the Lord's wrath be kindled against you, and he shut up the heaven, that there be no rain, and that the land yield not her fruit; and lest ye perish quickly from off the good land which the Lord giveth you." Deuteronomy 11:17

China is the third largest country and has the highest population of any country on the earth. Since it is so populous, natural disasters that occur in China are devastating. In the years from 1876 to 1879, very little water fell on the northern part of China causing a severe drought. The drought destroyed more than 300,000 square miles of crops and left millions of people and animals with no food. With the masses weakened by hunger, disease became rampant. The Chinese government tried to send food but had a difficult job determining which areas were priority. During transport, the carts and wagons of food were attacked by starving people. Even the animals pulling the loads were eaten. Hunger, sickness, and contaminated water killed thousands of people. Starving animals killed people. People killed each other over the little food that was available in certain areas. Nobody knows exactly how many people perished but the estimates are staggering. The experts say between nine and thirteen million peopled died.

China

At the same time, 1876 to 1879, a drought occurred in India also. The rains associated with the monsoons remained over the ocean never reaching land. The crops of Southern India failed and people starved. The next year, the drought hit Northern India and too much rain in the South destroyed the crops again! The government refused aid from other countries. Officials felt that borrowing from outside sources would put them into such severe debt that the country would suffer worse than the present distress. The government's isolationist policy cost five million people their lives.

India

Only eight years later in 1887, China's Yellow River flooded and destroyed thousands of acres of crops. Another severe famine was the result. This time, up to six million people died.

Drought occurs frequently in the United States. A heat wave followed by drought hit the Central and Eastern U.S. between June and September of 1980. Around 10,000 people died and economic losses were estimated at 20 billion dollars. During the summer months of 1988, the Central and Eastern U.S. suffered severe heat and drought again. The devastating result was 40 billion dollars in damages and between 5,000 and 10,000 deaths, including those related to heat stress. In the summer of 1993, the southeastern U.S. was the target of a heat wave and drought. The death toll was undetermined but the damages reached one billion dollars. From 1995 to 1996, a severe drought crippled the agricultural regions of Texas and Oklahoma. The crop losses in the southern plains cost four billion dollars.

The almost automatic association with drought in the United States is the 1930s. From 1930 to 1936, every state in the U.S. suffered from drought except Vermont and Maine. A severe drought for most of the 1930s in the Great Plains destroyed some 50 million acres of crops (mostly wheat and corn) in at least eight states. The devastation of this drought was multiplied by a drop in market prices (due to overproduction in 1931), a severe economic depression, insect plagues, and bad agricultural techniques.

The prairie grasses of the Great Plains had been plowed under or over grazed, which left the land vulnerable to the dry winds. One of the world's richest agricultural regions lost an estimated 350 million tons of valuable topsoil. The soil was lost to terrible dust storms called "black blizzards," which darkened the skies for hours or even days at a time. The dust storms started in 1933 and continued for years dubbing a region of Texas, Oklahoma, Kansas, Colorado and New Mexico as "the dust bowl".

Rags stuffed into cracks and wet blankets over windows and doors did little to keep the dust out of homes. Dust entered the smallest openings to settle on beds, dishes, clothes, floors, food, and people. The dirt and sand blasted paint off cars, machinery, and houses. Many animals that had not died from starvation, suffocated from dirt getting into their nostrils and lungs during the nasty storms. For years during and following the drought, people suffered from respiratory problems and other ailments caused by the flying dirt. The millions of tons of airborne dust also killed birds, closed schools and businesses, and blocked roads and railways.

To make matters worse, dry hot weather is perfect for hatching grasshoppers — and hatch they did! Millions of grasshoppers attacked the already stunted and dying crops. They ate 315 million dollars worth of crops between 1934 and 1938. They also ate the bark and leaves off trees and sometimes even the paint off houses. Even when they were killed by poisons, they were a nuisance. There were so many dead grasshoppers on the road that it became slippery and cars started sliding out of control. People raked huge piles of dead grasshoppers from their yards. Although the nickname "The Dirty Thirties" came from the dust storms, the grasshopper problem added validity to the title.

Hungry rabbits took their shot at the crops as well. Community members had huge rabbit rounding parties to drive the scoundrels into pens. Sometimes as many as 2000 people would surround a five mile area and drive the rabbits into the center where an open holding pen was waiting. Drives like this could capture as many as 6000 rabbits. The rabbits were killed and then eaten or sold as food.

Needless to say, the many problems of the Great Plains spelled out financial ruin for thousands of people. Whole populations dwindled as farmers and ranchers left to find jobs elsewhere. By 1936, around 50,000 people were leaving the Great Plains each month. Most of these people went to California looking for work only to meet thousands of others in the same predicament. From mid-1935 to the end of 1937, California's population increased by 221,000. At least 84 percent of those people came from the Dust Bowl region. In 1939 alone, Kansas lost 300,000 of its residents. The Great Depression was going on at the same time so there were very few jobs available and wages were very low. When the rains returned, some people stayed in their new locations and some returned to farming the Great Plains. The droughts had ended by 1939 — just in time for World War II.

What It Is

We always have the same amount of water. The earth never loses and never gains even a single drop of water.

Oceans, rivers, lakes, ponds, puddles, streams, animals, and plants send water skyward in the form of water vapor. At some point, the water condenses and is sent back to the ground as precipitation. The hydrological cycle, or water cycle, then starts over. Redistribution of the water can cause floods in one place and droughts in another. The water vapor does not always condense and fall on land where it is most needed.

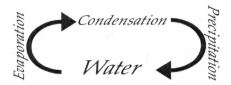

There is no universal definition of drought. Drought occurs in and affects many economic and social groups as well as geographic locations and political systems. Because each country or area that experiences drought has different impacts, different "normal" precipitation values, and different approaches and perspectives, definitions vary by region, group, and government. A simplified attempt to define drought may be: prolonged, drier than normal weather. (What is normal? What is prolonged?) The National Weather Service clarifies the general terms with a more specific definition. Drought is classified as 21 consecutive days of 30 percent (or less) of the average rainfall for a certain area and season.

Most of the time, we watch the weather forecasts hoping for sunny weather. When the forecast calls for rain, we grumble and complain. Our picnics, swimming parties, outdoor wedding receptions, hikes, camping trips, flea markets, sporting events, parades, open air concerts, fireworks displays, airplane trips, construction work, line-dried laundry, driving conditions, and marathon races are all spoiled by a rainy day. We curse the weatherman when he predicts rain and praise him when the forecast is fair skies. Drought is a natural disaster that we do not normally think about or notice until it is well upon us. It sneaks up on us in the midst of a long stretch of beautiful weather.

Drought can occur anywhere in the world. Every continent and region has experienced drought and is at risk for future droughts. One small area, a large region, a whole country, or even an entire continent can be hit by drought. It can affect us during any season of the year. It can last for weeks, months, or years. In short, drought can strike anywhere, anytime, and for any duration.

Norma Johnson

During a dry spell, West Clear Creek in Arizona was nothing more than a few scattered puddles.

How It Happens

The infinite number of variables that affect our weather make it impossible to determine the cause for every drought. Even when a cause is known, it is not known how other variables in our environment may have contributed. Droughts have been traced to interactions of wind, oceans, land, pollution, volcanic eruptions, and of course — the sun.

Even though weather is extremely complex, there are a couple of very simple concepts that help clarify some of the causes of drought. The first concept is pretty well known: warm air evaporates water and cool air condenses it. Since about 85 percent of moisture in the air comes from our oceans, the air over the oceans contains more water vapor than the air over the land. Luckily, the air is always circulating and trading places.

The second basic weather concept is also fairly well known: the higher the air rises, the cooler it gets. This is because the warmth we feel comes from the earth radiating the sun's heat. The sun's energy penetrates the earth which then radiates heat into the atmosphere. The sun's energy does not heat the atmosphere on the way down to earth.

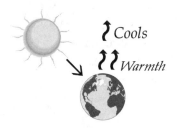

Further, the land absorbs and releases heat much faster than the water. That is because the sun's heat penetrates the thick soil only a few inches. The sun's heat can penetrate the water about 80 feet! The currents mix the water with cold deeper water which also slows the absorption and release of heat. During the day, the land (and the air over land) is warmer than the water (and the air over water). At night, the opposite is true. The land quickly loses its heat after the sun goes down, while the ocean's heat is released slowly. These differences affect the temperatures of the seasons as well. In the summer, the land is warmer than the ocean. In the winter, the ocean is warmer than the land.

With those two basic concepts in mind, let's discuss wind as a cause of drought. A change in either the force or the direction of the wind can result in drought for certain areas. Wind carries moisture-laden air over the land where it meets with cooler air. The water condenses and gravity eventually pulls the water down to land. If the winds do not blow with their ordinary force, the water could condense too early or too late to fall where it normally would. A change in wind direction can also cause precipitation to fall in the wrong place. Winds that blow from the ocean over the land carry with it the life-giving water that the land needs. If the winds shift and the moist air blows back toward the ocean, the water will probably end up back where it started instead of on the land.

The ocean is also a key ingredient in the making of a drought. Warm and cool ocean currents affect the air directly above. There are several cool ocean currents (coming from polar regions toward the equator) that run near land. The air above these currents is cooled and contains less water vapor. In addition, when the cool ocean air meets with warmer, drier land air, evaporation is increased as the ocean air is warmed. This exact phenomena is what creates some of the Peruvian and Chilean desert climates along the coast of South America. When warm ocean currents run near land, the warm air above is full of moisture. When it meets with the cooler land air, the moisture condenses and torrential rains result. The warm current that periodically replaces the cold water near South America has caused severe flooding in the normally dry areas. This warm current is called, "El Niño" and is known to affect the weather of several continents. (See the section on El Niño under *Floods*.)

The contour of the land also contributes to drought conditions. For example, it is common for mountains to have a moist fertile side and a dry infertile side. As warm air travels up the side of a mountain it cools, condenses and precipitates on that side and the top of the mountain. As the air travels down the other side, it warms and moisture is evaporated. Logically, many deserts are in the low lying areas in the basins of hills and mountains. High pressure helps to trap the warm air in the bowl shaped topography and prevent condensation. These are often normal climatic conditions resulting from the landscape and not necessarily a drought situation.

Speaking of pressure, atmospheric pressure also contributes to droughts. High pressure is heavy cool air and low pressure is light warm air. Meteorologists refer to areas of high and low pressures as simply highs and lows. A high usually means dry weather is in store and a low brings unstable to stormy weather. Highs and lows continuously replace each other and contribute to our ever changing weather. If a high pressure system is stalled somewhere, that area may experience drought. A low pressure system that is trapped behind it can cause too much rain for another area.

The sun plays a major role in the drought drama. First of all, the sun is the source of all our weather and affects every aspect of the conditions just discussed. Any cause of drought can be traced back through the web of environmental circumstances to the head honcho — the sun. If some of the sun's energy is blocked (by pollution or volcanic ash for example), the oceans can cool. When the ocean water is cooler, less evaporation takes place. Since most of our rain comes from evaporated ocean water, blocking some of the sun's energy can, in theory, cause droughts.

If the sun's heat is trapped in the atmosphere (by the greenhouse effect for example), global warming occurs. Scientists worry that the earth's temperature will increase faster than we can modify our existing behaviors to meet the challenges. No matter how complicated the weather patterns or how long the chain of environmental factors, the starting point is always the sun.

Norma Johnson

Long periods of dry weather become apparent in some of the vegetation of Canyon de Chelly, Arizona.

The sun also plays an important role in theories of drought cycles. By studying tree rings, scientists are able to look at the history of droughts. Trees grow a new ring each year. The size of the ring is related to the amount of rainfall for that year. The rings are much thinner during dry years and thicker during years with plenty of rain. By looking at very old trees, studying records, patterns of water levels, layers of sediment in river and lake beds, and other historical information, a history of droughts can be assimilated.

The history of drought is used by some to theorize a definite pattern that is related to sunspot activity. Sunspots are dark marks on the sun that are believed to be gases associated with the electromagnetic activity of the sun. These spots change in number each year but come and go in cycles. Drought cycles are thought to be associated with the cycle of sunspots. This is a theory that is being studied in hopes of predicting future droughts.

Other scientists say that there is no regular cycle of droughts. Either way, everyone agrees that drought does not follow any hard and fast rules and that it can strike at any time. Even if there is a notable drought cycle, it can rear its ugly head during "off times" and take us by surprise.

The Destruction

The hand of drought has many fingers. One of the devastating results of drought is famine. A drought in a farming area can affect the food supply of distant areas. In this way, drought can have far reaching consequences. Famine is an extreme shortage of food that lasts long enough to cause widespread hunger. Famines can have natural causes or man-made causes. Drought, floods, recurring hailstorms, insect plagues, and bizarre weather are examples of natural causes of famine. War, economic situations, disrupted imports, bad agricultural planning and techniques, poor distribution, and lack of preparation are a few examples of man-made famines. Whatever the cause, famine leads to starvation and death.

When people are undernourished, they become very susceptible to diseases and health problems. Drinking contaminated water during drought compounds the problem by introducing bacteria and additional sickness. During famines, many people die of starvation, but even more die from diseases that their weakened bodies cannot fight. Strong adults can recover from the effects of a short lived famine, but children and older people often have permanent health problems.

Drought sets up conditions favorable to many other types of disasters. Wildfires are common occurrences during times of dry weather (see chapter on *Wildfires*). Many fires are started by lightning or careless recreationalists but drought conditions worsen the devastation and facilitate the ignition. It is a common theory that the O'Leary's cow started the Great Chicago Fire but few realize that drought was a major factor.

Dust storms are common when the ground is dry, natural grasses have been plowed or grazed away and the land has been poorly managed. These horrible storms cause scoured skin, poor to zero visibility, damaged property, ruined machinery, loss of valuable topsoil, respiratory ailments and even suffocation.

Heat waves often accompany long periods of sunny weather that can add to the death toll of a drought. Many people experience heatstroke, heat exhaustion, sunburn, hyperthermia, and dehydration in times of drought and extreme heat (see *Severe Weather*). An increase in water consumption and electricity use are common during a drought coupled with a heat wave. These extra demands may cause resource shortages and government imposed restrictions. When drought causes rivers to become shallow or dry up, river transportation is disrupted and economic hardships are created.

What We Can Do

Since early history, people have been solving drought problems by irrigation and aqueducts. It is no different today. Irrigation is one of the main methods of fighting the effects of drought and bringing water to normally unfarmable land. Irrigation can be done by damming rivers which creates lakes called reservoirs. Canals and ditches bring water from the reservoir to places of need. Dams also help control flooding and convert water power to electric power.

Another irrigation technique is to tap groundwater. Much of the earth's water is trapped underground between rocks and in large spaces called aquifers. In fact, there is more fresh water underground than on the earth's surface. Scientists believe that there is thirty times more fresh water under the ground than in all the world's rivers and lakes combined. Wells tap into this subterraneous source for drinking and irrigation. When we use the water in underground pools faster than it can be refilled by nature, we have to drill deeper and deeper. Eventually this practice will deplete our valuable underground resource.

Another method we have for fighting drought is improving our agricultural practices. Plowing in contours that better hold rainwater and leaving some land idle allow moisture to build up in the soil. Planting cover crops during off seasons helps to enrich the soil and prevent dust storms. Regulations and practices that prevent overgrazing of the land also help to maintain topsoil value.

Man has also tried to control weather by inducing rain. From ancient civilizations to modern scientists, people believe that we can make it rain. Many societies punished the saints or rain gods of their particular superstition or religion in order to bully the rain out of them. Other societies created specific recipes for making rain clouds. Raindances, special prayers, inducing pity from Gods, and dressing up like clouds (sort of a decoy type practice) are more examples of our efforts to cause rain. Because rainstorms often follow battles, fireworks displays, and explosives testing, some theories arose that gunpowder produces rain. Experiments have proven disappointing and all but a few eccentric stragglers now believe that any link between gunpowder and rain is purely coincidental.

A modern practice of making rain has proven successful but benefits are controversial. The method is called cloud seeding and the "seeds" are chemicals. Originally dry ice, or solid carbon dioxide, and later silver iodide crystals were dumped or shot into clouds. The chemicals drop the temperature in clouds and ice crystals form and grow until they succumb to gravity. Basically, the chemicals turn the cloud into snow and the snow falls, melts and rain is the result. The possibilities of this technology include making rain, weakening hurricanes, and preventing hailstorms.

While we know that the process works, the practicality and usability are questionable. There are several flaws with the general idea. Obviously, clouds must be present to make use of this practice. Moreover, the clouds must be rain clouds. Some clouds are considered fair weather clouds which are not capable of producing rain even with chemical help. This causes two serious limitations. First, we have no way of moving clouds from one place to another and no way to create clouds. Therefore, droughts under clear blue skies (which is often the case) get absolutely no help from chemicals. Secondly, if rain clouds are present, how do we know it would not have rained anyway? We don't. There is no way to know if cloud seeding is actually useful or if it only initiates the inevitable.

Studies with cloud seeding have produced results that appear successful. Data shows that an increase in rainfall occurred. Since rainfall amounts are constantly changing and we have no way of knowing how much it will rain tomorrow, "increased rainfall" means nothing. It may very well have been a rainy season without man's efforts.

Let us assume that chemical cloud seeding actually does cause the rain to fall when it normally would not have. What happens to the area that would have received that rain? Drought? By inducing rain clouds to drop their load *here*, nothing is left for *there*. Changing nature in one place is bound to have effects in another. One area may gratefully reap the rewards of science, but it is likely to be at the expense of other areas.

The methods for fighting drought that we have discussed are either already in use, proven ineffective or still being studied. However, there are three things that we can do and should start doing today. First, individual families, organizations, church groups and governments need to store food and water. When drought strikes and famine results, we will be prepared. We will be in a position to offer relief to other families, other towns and other countries. By storing extra during times of plenty, we will ease some of the stress and worry of hard times. We will have food to eat and water to drink (see section 3).

The second thing we need to do is protect the environment. To guard against global warming we should become aware of environmental issues and their effects. Reducing pollution helps to protect our immediate surroundings, the atmosphere and fresh water supplies. Ending needless deforestation protects wildlife, prevents landsliding disasters and decreases the amount of carbon dioxide that is added to the atmosphere. This of course would help to lessen or slow down the greenhouse effect. By listening to environmentalists and initiating or supporting government attempts to protect our environment, we can improve the quality of our future.

The third thing we can do is conserve water now. When water shortages threaten, restrictions are placed on water usage. People may not be allowed to wash their cars or water lawns or even bathe daily. When the shortage crisis passes, we go back to using water liberally. We should continue to conserve water every day.

If every person saved one gallon of water a day, a four member family could save 1460 gallons of water in a year. The United States, with its population of more than 265 million, could save 96.7 billion gallons of water a year! Imagine if every person would save five gallons of water each day. The United States would conserve 483.6 billion gallons of precious fresh water in one year.

Saving five gallons of water a day is really very easy! Turning off the shower only one minute earlier than normal could save one to three gallons of water. Five gallons or more could be saved by bathing in a few inches less of water in the tub. Turning off the water while shaving in the shower or at the sink would also conserve water. Repairing any dripping faucets or hoses will save several gallons a day. Adjusting the water level in your toilet tank could save up to a half-gallon of water with each flush. Some tank mechanisms are adjustable but if yours is not, you could buy a new low-flow toilet or simply place a large heavy object (such as a plastic container filled with sand) in the tank to displace some of the water. Using buckets of soapy water to wash vehicles rather than a running hose will reduce water use by more than five gallons. Use your imagination and you will see that conserving five gallons of water a day is no hardship.

Check your water bill this month and make a goal with your family for each person to conserve five gallons each day. Compare the gallons used with the next bill and see what a difference you have made. You will also save money that your family can use for something more fun than a long shower!

Earthquakes and Tsunami

"...and there was a great earthquake, such as was not since men were upon the earth, so mighty an earthquake, and so great."
Revelation 16:18

In October 1905, The National Board of Fire Underwriters made its report public. That report said, "San Francisco has violated all underwriting traditions and precedents by not burning up. That it has not already done so is largely due to the vigilance of the Fire Department, which cannot be relied upon indefinitely to stave off the inevitable." [Thomas, G. and Witt, M.]

Six months later, San Francisco was destroyed by three days of uncontrolled fire that followed the strongest earthquake in California's recorded history. Unfortunately, the expertise of the chief of the city's fire department could not be utilized. He died from a falling chimney in the earthquake.

On April 18, 1906 at 5:13 A.M., 250 feet of the San Andreas Fault shifted causing damage to an area 450 miles long and 50 miles wide. The tremor jarred, buckled, and shook up the great city of San Francisco. Buildings crumbled, gas and water mains broke, rail lines twisted and deformed, glass shattered, streets split open, and familiar landmarks disappeared. The structures built on filled land experienced the worst damage, collapsing like cards.

The broken gas and electric lines sparked and fed fires all over the city. The broken water mains left firemen frustrated and helpless as they tried to fight the fires. They had to resort to artillery and explosives to try to keep the flames in check. In some instances, this technique actually made the fires spread. The city burned on for 72 hours causing even more damage than the earthquake.

In the end, the disaster destroyed 28,000 buildings in San Francisco alone and left over 300,000 residents homeless. The economic damage was estimated between 300 and 500 million dollars and the death toll around 700. The number of injured is believed to be over 5000. The city rebuilt itself on the same unstable soil and even filled in some of the Bay with the rubble from the earthquake and built on top of it. Given the surety of future quakes in this area, that was a dangerous gamble.

California

San Fransisco

It would be nice for the rest of the world to feel secure and safe from earthquakes since we do not reside in California. Unfortunately, we are all at risk. In fact, the state with the most major earthquakes is not California. It is Alaska. One of the largest earthquakes ever recorded at a magnitude of 9.2 on the Richter scale (U.S. Geological Survey National Earthquake Information Center) occurred near Anchorage on Good Friday, March 27, 1964. Over 200,000 square miles of the earth's surface suffered vertical displacement. (That's about the size of Illinois and Indiana together!) In some places, the ground lifted 30 feet. The 50 foot rise of the sea bed caused huge tsunami that killed 122 people. The tremendous waves destroyed towns along the Gulf of Alaska and on the western shores of The United States and Canada. Landslides and avalanches caused more destruction. The damage from this earthquake amounted to an estimated 500 million dollars!

Alaska

Anchorage

The New Madrid region of Missouri experienced extremely powerful earthquakes in 1811 and 1812. There was no measuring equipment at that time but recent scientists have estimated that several of the shocks were greater than 8.0 on the Richter scale. Vibrations were felt in Washington, D.C., Virginia, New York and Boston, more than a thousand miles distant. The death toll is estimated to be very low because of the sparse population at that time. Now the population has grown, the threat of a major earthquake remains and the memory or knowledge of the past has faded. Unlike Californians, the people of Missouri may not be as prepared for or concerned with the very real threat of destruction just under their feet.

Missouri

New Madrid

Every state in America, every country in the world and every person on earth is at risk for earthquakes and should prepare for the possibility. (Although some places have a much higher risk than others.) Millions of earthquakes occur every year on the earth, thousands of them are felt by people and hundreds of them can cause injury, death, and destruction. According to The National Earthquake Information Center, every year the world averages more than 900 earthquakes of magnitude 5.0 or greater.

G.K. Gilbert/USGS

San Fransisco Earthquake, 1906. East side of Howard Street near Seventeenth Street. All houses shifted toward the left. The tall house dropped from its south foundation wall and leaned against its neighbor.

G.K. Gilbert/USGS
Damage on Shotwell Street, San Fransisco, from the 1906 earthquake.

Mendenhall/USGS
The Agassiz statue at Stanford, April 21, 1906.

Mendenhall/USGS
View from the corner of Geary and Mason streets, S.F., 1906.

Mendenhall/USGS
Nob Hill, S.F., from corner of Van Ness and Washington Streets, 1906.

American Red Cross
View of the fires caused by the 1906 San Fransisco earthquake.

U.S. Army/USGS
The Alaska earthquake of March 27, 1964, caused the collapse of Fourth Avenue near C Street in Anchorage.

What They Are

Any measurable movement or vibration of the earth's surface is called a temblor or seismic event. These events are more commonly referred to as earthquakes and usually occur along fault lines. (It is important to note, however, that an earthquake can occur anywhere.) A fault line is a crack in the surface of the earth where the two sides move with relation to each other. (If the two sides of the crack or fault are not moving, the fault is said to be dormant.) Since the walls of these cracks are not smooth, the rock masses and earth materials stick together and build up pressure. When the forces of movement overcome the binding friction of the rocks, the pressure is suddenly released and an earthquake is the result.

The point underground where an earthquake originates is known as the hypocenter or focus. The point on the surface directly above the hypocenter is known as the epicenter. Figuring out where an earthquake is centered is fairly easy. By measuring the direction of the vibrations and time of arrival at two or three widely spaced measuring stations, the data can be plugged in to specific equations. The equations determine the distance to the epicenter from each measuring station.

The vibrations from an earthquake travel great distances and are known as waves. Earthquakes generate several different kinds of waves through the surrounding ground, including S-waves and P-waves. S-waves travel along the surface in ripples much like ripples in a pond. If you were to look at an S-wave from the side, it follows the up and down pattern of a typical sine wave. By stretching a rope across a room, attaching one end to something and shaking the other end up and down, you can see the artificial S-waves move across the rope. Many eyewitnesses to earthquakes have seen the rippling S-waves move along streets, sidewalks, or open fields. These characteristics give S-waves the names surface waves and sinusoidal waves. Another name for an S-wave is shear waves. This is due to the motion of the ground at right angles to the direction of the wave. In general, S-waves cause the side to side or up and down shaking force that topples buildings, chimneys, and other structures.

Pressure waves, or P-waves, are not easily seen along the surface. They represent a compression of the earth's rocks and other materials transferred from particle to particle. It is hard to imagine rocks and other hard materials compressing but with the forces of an earthquake, they do. They act like a spring. If you secure one end of a coil spring, then pull the free end and release it, you will see the compression wave along the spring coils as they come together and then bounce apart again. The same thing happens to the earth. These pressure waves travel thousands of miles and penetrate deep into the earth. Scientists have been able to gain a greater understanding of the earth's interior by analyzing P-waves and their differing velocities through different materials.

How They Happen

There are several things that can trigger an earthquake. For example, atomic testing and underground explosions will often cause an earthquake. Many times when a dam is built on a river, the weight of the resulting lake that is formed causes rocks to shift and buckle and earthquakes result. It has been discovered that when large amounts of water seep into the earth, the liquid may act as a lubricant which allows rocks to slide and triggers earthquakes. Scientists have found that by injecting water and other liquids into the ground under high pressure, they can induce slippage of rocks previously held together by friction. There is some question as to the benefits of inducing several small earth-quakes along fault lines that seem to be "stuck" rather than waiting for the energy to build up and cause larger earth-quakes. Theoretically, this sounds like a good idea. However, it is too dangerous to test this method on large faults and in populated areas where the most benefit lies. The risk of setting off a large earthquake is too great.

The majority of earthquakes are caused by the slow, yet continual, movement of the earth's surface. The basis of all modern geology is a theory called *plate tectonics*. According to plate tectonics, the earth's crust is broken into several pieces or plates that float on the layer below, like icebergs float in water. These crustal plates vary in thickness from about five miles to fifty miles. This "cracked egg shell" crust is not stationary but continually moving. As the edges of the plates push against one another, many interesting things happen, including earthquakes.

Most of the world's major earthquakes occur along crustal plate boundaries. As plates slide alongside one another or dip underneath and thrust over the top of one another they get stuck and build up stress before slipping. Mountain ranges are formed by the crustal plates buckling and rock and earth being pushed upward as two plates push against one another. Scientists say that without the constant movement of the earth's crust (earthquakes), our planet would be as smooth as a polished cue ball. Although that may or may not be true, earthquakes are an important and natural part of the earth's constant change and evolution. They have also given us a window into the workings of the interior of the earth. More than any other source, earth-quakes give us information about the structure and functioning of the inner planet.

How They Are Measured

The shaking of the earth is recorded on measuring devices around the world called seismographs. These instruments measure time in seconds on the horizontal axis and ground motion in millimeters on the vertical axis. The purpose is to record on paper or magnetic tape the vibration of the earth's surface. The paper or tape copy of the recording is called a seismogram, and the component of the measuring device that jiggles when the earth moves is called the seismometer. This is usually a heavy pendulum or block mounted on a spring.

Determining the severity of an earthquake is another story. Most people are accustomed to hearing the magnitude as measured on the Richter scale or the intensity as measured on the Mercalli scale. The Richter scale is not an instrument or physical scale that measures magnitude like a weighing scale measures pounds or a seismograph measures vibrations. After Dr. Charles Richter, a geologist at California Institute of Technology, proposed his scale in 1935, many people wanted a look at this mysterious device. His widely accepted and used scale is actually a mathematical (logarithmic) scale measuring the magnitude of ground movement in microns. The scale starts at zero and is open ended.

Since earthquake waves tend to weaken with distance, the magnitude of an earthquake to a person right at the epicenter would be different than to a person miles away. The Richter scale sets the standard distance for measuring magnitude at 100 kilometers (about 60 miles) from the focus of the earthquake. Unfortunately, there is rarely a seismograph at exactly 100 kilometers from an earthquake so corrections factors are used to determine what the magnitude would be at that distance. This correction factor introduces some error into the measurement. Since all seismographs are not the same and do not measure vibrations in exactly the same way, there may be more correction factors introduced into the equation. These too produce some error.

For these reasons, it is common for two measuring stations at different locations to give different Richter magnitudes. The Good Friday earthquake in Alaska discussed earlier had reported magnitudes on the Richter scale from 8.2 to 9.2. The earthquake in New Madrid, Missouri on February 12, 1812 has been estimated as low as 7.7 and as high as 8.8. Although variations exist for virtually every earthquake, in general the Richter scale is a reliable estimate of the magnitude.

For those of us who are not scientists and do not have access to sophisticated measuring equipment, there is another scale. The Mercalli scale is not as precise as the Richter scale, but it is easier to use. The scale was first proposed by Giuseppe Mercalli in 1902. In 1931 the scale was modified and generally adopted by world seismologists as a standard scale. Because of the changes, the scale is referred to as the Modified Mercalli scale.

The Modified Mercalli scale is basically a rating, from I to XII, of the observed effects of an earthquake. If it is felt by only a very few people who happen to be on upper floors of tall buildings, the rating may be a I or II. If the shaking is strong enough to knock items off shelves and awaken sleepers, it would be a V. When chimneys and smoke stacks are toppled, unsecured houses slide off the foundation and are toppled, unsecured houses slide off the foundation and the shaking is felt by everyone, it would be an VIII. If there is total destruction, the Mercalli intensity would be XI or XII. The scale is actually very detailed so that specific effects are rated with specific intensities. Roman numerals are used to avoid confusion with the numbers of the Richter scale. Obviously the subjectivity of the observer may cause different intensities to be assigned to the same earthquake.

The Destruction

There are several contributing circumstances that determine the amount of destruction an earthquake generates. Magnitude is certainly a major factor, but it is not the only consideration. A magnitude 8 or 9 earthquake in a sparsely populated desert would cause less damage than a magnitude 5 or 6 earthquake in a major city. Location, therefore, is also a primary factor in the damage tally. Along with the geographic location, the depth of the focus is also important. In general, the deeper the earthquake, the less devastating the consequences.

Another element of the earthquake to consider is the duration. Some temblors last only seconds while others last several minutes. Logically, the shorter duration earthquakes usually cause less damage.

The time of day when the earth suddenly decides to move can mean the difference between life and death. The damage to structures would most likely be the same no matter when the earthquake strikes. However, the injuries and death tolls vary greatly with the time of arrival. If most people are still at home and indoors, there are generally fewer injuries and deaths. When people are out on the sidewalks, driving on the roads, and buzzing around the business districts, it is more likely that they will be injured by crumbling buildings, collapsing roadways, traffic accidents, etc. When office buildings, schools, stores, and other public facilities are not filled with people, their structural damages are purely economic losses and contribute little to the loss of life.

Obviously the construction of buildings plays a significant role in the havoc of an earthquake. Rigid structures such as adobe, stone, brick and other masonry types do not fare well when the ground underneath begins to move. They tend to crumble easily and their heavy materials are dangerous to those around. Flexible buildings such as steel frames, wooden homes, and steel reinforced concrete do much better as they sway rather than break. If these flexible structures are built too close together, however, they can swing into each other and act as battering rams! There are many building codes and regulations which are being enacted more and more to help avoid some of the terrible results of earthquakes.

The geology of the land is extremely important to the safety of buildings and their occupants. It has been seen over and over in history that the worst damage occurs to structures built on loose soil and landfills. (Even in Biblical times, the wise man built on a rock and the foolish man built on the sand!) A building can be built with the best methods and materials but if the land underneath sinks or liquefies, there is little chance of salvage.

When a house is built on granular material such as soil, sand, or sediment that has not consolidated, something called liquefaction can occur. When there is water in this granular material, it fills tiny pores and openings in between the grains of sand or soil. Normally the water does not exert enough pressure on the grains to push them out of place. When the S-waves of an earthquake travel through unconsolidated material, it raises the water pressure in the tiny spaces and the soil becomes an unstable, watery mess.

In the book *Superquake!*, David Ritchie suggests an experiment to demonstrate liquefaction. Fill a sandwich bag with water, close it, and set it in the sink. The water pressure inside the bag is not high enough to break the plastic. Then hit the bag hard with your fist. This will significantly increase the water pressure inside the bag and it will burst. The watery mess in the sink is akin to the watery mess of the soil when the shock waves smack the pores of water in the soil. Liquefaction is the most dangerous to moist, sandy soils that have been recently deposited. Soil that can hold large amounts of water and that contains very few boulders, rocks, or gravel is at risk for liquefaction.

Once the soil turns into a soupy slush, the oscillations of the S-waves become amplified causing severe damage. Drop a heavy object on the floor (consolidated material) and then do the same on a waterbed (unconsolidated material) and see which one makes the most vibrations.

Another aspect of the geology of the land is the distance the shock waves of an earthquake can travel. Some materials of the earth act as a damper and weaken the waves as they travel. Certain earthquakes are felt thousands of miles away while others are confined to a local region. This depends largely on the type of rock and earth that the earthquake waves encounter on their journey outward.

Earthquakes are particularly devastating because they incite other major disasters. Besides crumbled buildings, collapsed roadways, land fissures, and downed electric lines and water mains, there are additional dangers. Earthquakes routinely cause fires, landslides, floods, volcanic eruptions, and tsunami. (See corresponding chapters for more information.)

There is currently no method of consistently predicting earthquakes so they usually strike with no warning. Scientists around the world are studying precursors and scrutinizing characteristics of earthquakes in hopes of successful forecasting. Some earthquakes have been accurately predicted and researchers look optimistically to the future.

Even with a proven method of earthquake predictions, evacuations could be cumbersome and ineffective. If people are evacuated unnecessarily too many times, will they start ignoring the warnings? If there is a real threat, how do we evacuate millions of people in a short amount of time? Will only the wealthy be able to pick up and leave home and business at every signal? Will the general population react orderly when death and destruction threatens? Who should be evacuated first? These are just some of the concerns that will need to be addressed in the future.

What We Can Do

Since we cannot currently predict, avoid, stop, or control earthquakes, the best we can do is prepare for them. We can start by preparing our homes. If they are not already built to withstand earthquakes, there are a few things we can do to make them a little safer. One of the first things to fall in an earthquake is the chimney. For this reason, it is wise to brace the chimney and reinforce the ceiling near chimneys. Bracing the ceiling joists near the chimney is also a good idea. Anchor the floor joists to the foundation and porches and decks to the main house. Secure the water heater to a wall and plumbing and gas lines to walls, ceilings, or floors. Put flexible hoses on all gas appliances.

Check your insurance to see if you are covered for earthquake damage. It is likely that you need to purchase separate earthquake insurance or pay more for earthquake coverage. Upgrade your home to meet with all the current building standards in your area. When buying a new home, look for one that meets building codes and standards and one that is built on rock. A geotechnical expert can advise you about the type of land under the house. The county geologist or planning and development department in your area can tell you which areas are built on landfills, soil, sediment, rock, etc. Since houses built on loose soil and landfill are much more susceptible to earthquake damage, it is worth your time, effort and money to buy a house that is built on consolidated ground.

An important part of preparing for an earthquake is to have a family plan (see Section 2). As you go through your house looking for potential dangers, pay attention to the height and position of heavy objects. Heavy mirrors or framed pictures should not hang directly above beds, chairs, or sofas. Poisons, pesticides, fuel, etc., should be stored in a closed cabinet on the bottom shelf. Keep the cabinet locked if you have small children. Breakable items should also be stored low. Have earthquake drills with your family at different times of the day and night. Make sure you have a good first-aid kit, emergency supplies and extra food and water (see Sections 2 and 3).

If you are outside when an earthquake strikes, move to an open area away from buildings, trees and power lines. Many people have heard so often to stand in a doorway in case of earthquakes that they will run into buildings from the outdoors. Unless you are already very near a doorway, this could be a fatal mistake. The most dangerous place to be in an earthquake is just outside a building where many victims are buried by falling rubble.

If you are driving, pull off the road and stay in your car. Once the shaking stops, proceed with caution but avoid bridges and ramps. If you are inside your home or a public building, take cover under a desk, table, or other sturdy protection (including doorways or interior walls). Stay away from exterior walls, windows and other glass that could shatter. Remember that most deaths and injuries in an earthquake are due to falling debris and flying glass. Cover yourself as best you can! Do not try to go outside until the shaking has stopped. If you are near an ocean, lake or river, move to higher ground immediately.

Before you venture out to survey the damage to your property and neighborhood, make sure everyone in your family has on sturdy shoes and clothes. This is important as broken glass is a common and abundant post shock danger. If the damage to your home is severe, after shocks could cause further damage and collapse. In that case, you do not want to go in and out for things you need. If you have an evacuation kit near the entrance to your home, this should not pose a problem as you will have everything you need within reach. (Of course anything can happen in an earthquake and such luxuries as shoes and evacuation kits may be impossible to locate!)

If the damage is less severe or not so apparent, you will have to make a closer inspection. Check the entire length of the chimney as any damage could lead to fire. Check gas lines, electric wires and plumbing for leaks, breaks, and system damage. Gas leaks can cause fires and carbon monoxide poisoning so it is important to get outside. Turn off the gas at the outside main valve and call the gas company from an outside phone.

If you suspect damage to electric wires due to the presence of sparks, frayed broken wires, or the smell of hot insulation, turn off the electricity at the main fuse box or circuit breaker. If water pipes are damaged, contact the water company and avoid drinking tap water. Sewer line damage can also contaminate drinking water. Call a plumber, avoid using toilets, and only use water from your water supply (see section 3). If a plumber is not available due to the high demand following a disaster, use the methods for disposal recommended by your health department. Use drain plugs in sinks, bathtubs, and showers to prevent sewage backup.

Clean up any spilled chemicals, flammable liquids, or broken glass in the house and open cupboards and closets cautiously. Tie a red, yellow, or white signal ribbon on the outside of your front door. (See "Family Plan" in Section 2 for an explanation of colors.) Evacuate your home if you are at risk for flooding, fires, tsunami, or landslides.

After taking care of immediate dangers, you should help your neighbors with rescue efforts and first aid where appropriate. Do not enter damaged buildings and do not move seriously injured people unless further injury is likely.

Try to open roads for rescue vehicles and keep phone lines free for emergencies. Use caution around animals since behaviors can change dramatically during and after an earthquake. Listen to radio broadcasts for updates and emergency information.

Earthquakes are unpredictable and give little or no warning before striking. They can occur anyplace, during any season, at any time of day or night with a variety of magnitudes. By following the guidelines given, you can maximize your chances for safety during and following an earthquake. Since there is a lot if information in this chapter, the following is a summary of things to do before, during and after an earthquake.

1. Buy a home that is earthquake resistant and built on consolidated ground, or make structural repairs and improvements.

2. Have a good family plan, emergency supplies, evacuation kits, and extra food and water. Have drills with your family at different times of the day and night.

3. Go through your house securing bookshelves, and rearranging cabinets, wall hangings and furniture as necessary to minimize dangers.

4. Protect yourself during the shaking from falling debris and flying glass. If you are outside, stay outside and in the open. If you are inside, stay inside and under a sturdy desk or table.

5. Once the shaking stops, treat injuries and take care of immediate dangers such as small fires, leaking gas, and spilled flammables. Listen to a radio for updates and evacuate if necessary.

6. Help friends and neighbors with rescue efforts and neighborhood safety tasks such as opening blocked roads.

7. Make a full inspection of your home and property for structural damage and possible dangers.

8. Make repairs as necessary and start stocking up for the next earthquake!

Tsunami

A tsunami is a wave or series of waves that are mostly caused by undersea earthquakes. Volcanoes, landslides, and explosions near or under water can also incite tsunami. The word *tsunami* (both singular and plural) is Japanese and literally means "harbor waves" but is used to mean "seismic sea waves". These waves are sometimes called tidal waves but have nothing to do with tides or gravitational pulls.

When the ocean floor is displaced, the water above rises respectively, since water is noncompressible. The distur-bance sends ripples outward in rings like a pebble thrown into a pond. In deep water, these sea waves are only a few feet high and travel in excess of 500 miles per hour. (Regular wind-driven waves only reach speeds of around 60 miles per hour, even during storms.) When the waves reach shallower water, they slow down. The faster waves behind pile up and create a wall of water that can reach heights of over 100 feet.

Tsunami can travel across entire oceans and create havoc on several coasts. (Traveling at 500 miles an hour, sea waves still need at least 22 hours to cross the Pacific Ocean!) These seismic waves have been known to slosh back and forth across the ocean for a week. The crest to crest distance of the waves averages about 100 miles and sometimes successive waves are larger than the first. The Pacific basin is especially vulnerable to tsunami because of the frequent earthquakes there. In the United States, California, Oregon, Washington, Hawaii, and Alaska are at risk for tsunami.

Since seismic waves are relatively small in deep water, they are not detected by ships or aircraft. In order to prevent a surprise arrival on populated coasts, a warning system was established. In 1948, the Seismic Sea Wave Warning System was put into operation and was later strengthened by the addition of other groups. The Pacific Tsunami Warning Center is located in Honolulu and is administered by the U.S. National Weather Service. They monitor earthquakes in the Pacific and watch tide gauges. If an earthquake is likely to cause tsunami, a watch is issued to all the member nations in the Pacific. When actual tsunami are detected or observed, the watch is changed to a warning.

Tsunami are extremely powerful and can reach inland for several miles. They tend to scour the land as they crash and then recede. Some tsunami are preceded by a withdrawal of the water from the coast. It may seem like the ocean sucks the water from the shore and then spits it back from great heights. A tsunami in 1771 tore huge coral heads from reefs along the Ryukyu Islands in Japan. The massive coral, weighing up to 750 tons, was thrown several miles inland where it remains today. In addition, the main wave killed 11,000 people in the Yaeyama Islands.

Another tsunami in Japan, this time in 1896, brought about massive death and destruction. Three hundred miles north of Tokyo, on the coast of Honshu, thousands of people were gathered for the annual Shinto festival. A little less than an hour after a small underwater earthquake took place nearby, a wall of water 110 feet high hit the town of Sanriku. The powerful wave killed over 27,000 people, injured 9000 more and destroyed 13,000 houses.

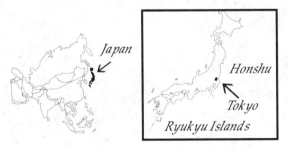

Damaging tsunami occur in Alaska and Hawaii about every seven years and even less frequently on the west coast of the United States. Although rare, tsunami have been responsible for more than 470 deaths and several hundred million dollars in property damage in the United States.

R.W. Lemke/USGS
A fishing boat beached several hundred feet inland from the head of Resurrection Bay due to tsunami from the Alaska earthquake of 1964.

U.S. Army/USGS
Wrecked boats and other debris at Resurrection Bay caused from tsunami resulting from the 1964 Alaska earthquake.

U.S. Army/USGS
Tsunami damage in the railroad marshalling yards in the Alaska Gulf region, 1964.

Andrea Booher/FEMA
This farm enjoys the benefits of floodplain location but suffers the consequences of nature's wrath.

Floods

"And the flood was forty days upon the earth; and the waters increased, and bare up the ark, and it was lift up above the earth." Genesis 7:17

Water is one of the world's greatest marvels. We drink it, we swim in it, we wash with it, we dream about it, we study it, we build houses with a view of it, we pray for it, we fear it. One thing we cannot do is control it. When we try to hold back the forces of water, we can exaggerate its power and magnify its potential. The survivors of a catastrophe in 1889 can attest to our illusion of dominion over water.

The South Fork of the Conemaugh River in Pennsylvania was dammed in 1852 to supply water to a nearby canal. Within ten years, the canal was replaced with a railroad and the dam's purpose became obsolete. In 1879, the dam became the private property of the South Fork Fishing and Hunting Club, an exclusive clique of millionaires. The dam held back Lake Conemaugh which was stocked with fish for the recreation of the 66 affluent members.

Over the next ten years, bad decisions and poor maintenance by the owners contributed to disaster. A meshwork cover was placed over the spillway to prevent fish from escaping the lake behind the dam. Pipes that allowed independent regulation of water levels were removed to avoid repair and maintenance costs. The breast of the dam was allowed to sink lower than the shoulders. The rich fishermen neglected to make needed repairs and carry out routine upkeep. Unfortunately, the industrial area of Johnstown, located 14 miles downstream, paid the price.

The month of May in 1889 saw record rainfall in the Conemaugh River valley. Several days of showers that month drenched the land and elevated the creeks and rivers. The downpour of May 30th concerned some of the inhabitants of Johnstown and other villages downstream from the South Fork dam. The voluntary evacuation of many of the observant locals saved their lives. Those who ignored the signs or were unaware of impending danger were not so fortunate.

Debris from the flooding rivers clogged the meshwork cover over the dam's spillway. Lake Conemaugh rose that day and the next until it spilled over the breast of the dam. A few minutes after 3:00 P.M. on May 31, the dam exploded. The 3 mile (5km) long and 1.2 mile (2km) wide lake which had reached a height of over 70 feet at the dam came crashing through the dam and drained on the cities below. It took only 36 minutes to empty the lake. Twenty million tons of water poured out in a 40 foot high wave that was half of a mile wide. It gushed downstream at an average of 40 miles per hour to inundate the 30,000 residents below.

As the churning water made its way downstream, it swept through the villages of Mineral Point, East Conemaugh, Woodvale, and others. Heading for Johnstown, the massive water picked up trees, houses, barns, animals, humans, railcars, boilers, and miles of tangled barbed wire. When the immense wave of water and debris hit Johnstown, it destroyed nearly everything in its path. Stone Bridge, at the lower end of town, finally stalled the rolling collection of water and accumulated wreckage.

Although the bridge saved most of the structures farther downstream, it did not spell relief from the disaster. The congested pileup not only sent a rebounding wave of water up the tributary of Stony Creek to demolish homes in the city of Kernville, it also caught fire. The ruins, fueled by oil from furnaces and railcars, burned unrestrained for days. When the flames finally died out, relief workers had to use explosives in order to clear the charred jumble at the bridge.

In the end, 2,209 people lost their lives, with another 967 missing and numerous homeless. Of the 2,209 known dead, there were 99 whole families, 396 children under the age of ten, and 777 unidentified victims. It is possible that many of the 967 missing may have been burned in the grisly fire at Stone Bridge. Over 1600 homes and 280 businesses were destroyed with property damages reaching 17 million dollars.

Pennsylvania   *Johnstown*

Was this a natural disaster or a man made accident? It is most likely a combination of both, but the Pittsburgh courts ruled otherwise. They decreed the catastrophe to be an act of God. The members of the South Fork Hunting and Fishing Club were not legally responsible for the damage or loss of life and, therefore, did not suffer any financial repercussions. Some of the members did offer relief contributions but none of the legal suits against them ever paid. A widely quoted poem by Isaac Reed illustrates a popular attitude toward the disastrous event:

> *Many thousand human lives —*
> *Butchered husbands, slaughtered wives,*
>   *Mangled daughters, bleeding sons,*
>   *Hosts of martyred little ones,*
> *(Worse than Herod's awful crime)*
> *Sent to heaven before their time;*
>   *Lovers burnt and sweethearts drowned,*
>   *Darlings lost but never found!*
> *All the horrors that hell could wish,*
> *Such was the price that was paid for — fish!*

American Red Cross

*Johnstown, PA on May 31, 1889. An artists rendition of the
tremendous devastation wrought by the raging flood.*

American Red Cross

The west end of Main Street in Johnstown, PA following the flood of 1889.

The ghastly event at Johnstown could easily replay itself in another location in the United States. A report from the American Society of Civil Engineers documents the dangerous situation of dams in the United States. Summarizing information from the Association of State Dam Safety Officials and the U.S. Army Corps of Engineers, the report paints a scary picture. There have been over 200 dam failures in the United States in only the last ten years. Since 1990, damage and repair costs for dam failures in just eight states have exceeded 60 million dollars!

One fourth of all U.S. dams are older than 50 years. Most older dams are built with inadequate spillways which is a major cause of dam failure. Of the 75,300 regulated dams in the country, 9200 are considered high-hazard (failure would likely cause loss of life and significant property damage), and 2100 are considered to be unsafe. Thirty-five percent of high-hazard dams have not been inspected since 1990. In addition, many dams that are more than 50 years old are abandoned and have no known owners. These dams are not regulated or inventoried. They are not inspected and nobody is paying for repairs.

Every state has at least one dam that is considered high-hazard. Pennsylvania and Texas each have over 500 high-hazard dams and North Carolina has the most with 874! The 1995-96 National Inventory of Dams revealed that a majority of high-hazard or significant-hazard dams have no Emergency Action Plans in place. Responsibility lies mainly with state governments for the regulation of dams, but federal funding is necessary to improve or initiate dam safety programs.

Besides dam failures, river flooding is prevalent throughout the world. The Yellow River (Hwang Ho) in China floods regularly causing damage to crops and property and extensive loss of life. In 1887, the river flooded so badly that over 900,000 people died! The water topped levees that were 70 feet high and flooded over 300 villages. Twenty to thirty feet of silt laden water covered some 50,000 square miles of crop land and forced over two million people from their homes.

Bangladesh is another densely populated area that is prone to frequent flooding. It is actually the most densely populated country in the world with 119 million people crammed into an area smaller than Wisconsin. The country is on the coast of the Bay of Bengal which is a common site for cyclones. Most of the country is only slightly above sea level and is a large flat plain cut into many islands by hundreds of rivers and streams. The rivers flood regularly (due in part to the rainy climate) and with nowhere to evacuate, the results are usually disastrous. If the flooding rivers coincide with high tides and/or tropical storms, the outcome can be cataclysmic. In the last century, all but a handful of the world's most disastrous floods (claiming over 10,000 lives) have occurred in what is now known as Bangladesh.

Every part of the United States is threatened with flooding and during recent years, several states have suffered major disasters. A look at the news for the past few years reveals some of the worst flooding in history over various regions. In 1997, the Red River flooded parts of North Dakota, South Dakota, Minnesota, and Canada driving thousands from their homes. The city of Grand Forks, North Dakota, with a population of 52,000, was especially hard hit.

North Dakota

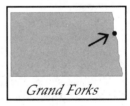

Grand Forks

Grand Forks had experienced one of its worst winters that season (1996-1997). Eight major snow storms dumped 98.6 inches of snow on the city. The National Weather Service predicted that a major flood was likely to occur due to the spring melt. They predicted a crest of 49 feet at Grand Forks which is well above the 28 foot flood stage. The dikes were raised from 50 to 52 feet to try to contain the water and save the city.

The river reached its 28 foot flood stage on April 4, while the last of the eight blizzards was just getting underway. The Red River, continuing to rise, topped the dikes on April 19.

It finally reached its crest at Grand Forks on April 21. Unfortunately, the predictions undershot the high point by more than five feet. The river crested at 54.11 feet and immersed the Red River Valley. Thirty-nine thousand people were evacuated with the few possessions they could grab.

When the river receded, the work began. The flood waters left behind so much destruction, dirt, and muck that clean up was a major undertaking. Almost everything that the waters soaked had to be discarded. Walls had to be torn apart, sheet rock, insulation, and all, before rebuilding could begin. The flood's resulting rubbish totaled 224 million tons, nine months worth of normal garbage.

Andrea Booher/FEMA

The Red River flood in 1997 caused this camper in Grand Forks to sit so out of place.

Andrea Booher/FEMA

These two people survey the neighborhood in Grand Forks, ND following the 1997 Red River flood.

What They Are

In general, a flood is a surplus of water on land that is normally being utilized by man. When the soil, vegetation, and atmosphere can no longer absorb the excess water, it accumulates on the land. Floods can also be an overflow or spillage from any body of water onto bordering land. River flooding occurs when the waters of rivers, streams, brooks, and creeks spill over the natural banks onto the neighboring plains.

The land that borders rivers is usually very fertile, good for transportation, commerce, and waste disposal, and is appealing. River valleys are therefore normally populated, and are known as the floodplains. Government regularly maps floodplains for insurance purposes and for developing regulations for floodplain management. Any land that has a one percent probability of flooding each year is known as the "100-year floodplain". In other words, this land will likely experience flooding at least once every 100 years.

Andrea Booher/FEMA

The Midwest floods of 1993 left motorists stranded and frustrated.

The Federal Insurance Administration spends over 36 million dollars annually to publish or update flood risk information. They get support from state and local governments as well as several specialized agencies. It is known where flooding is likely to occur. We continue to build on these flood-prone banks because of the important resources and enticing benefits of riverside communities. The frequent occurrence and impact of recent devastating floods will support continued and improved floodplain management.

Coastal flooding is oceans and seas extending onto land that is being put to use. This usually occurs because of tropical storms, extra high tides, high winds, other types of severe weather, and tsunami.

Flash flooding is sudden and unpredictable surges of water, usually due to heavy localized rainfall. Many natural disasters also cause or contribute to sudden flooding. Flash floods are very dangerous because they are not anticipated and catch people in jeopardous situations. A high percentage (some reports say as much as 80%) of flood related deaths occur to people driving in automobiles. Finding motorists who are unsuspecting, unprepared, and/or uninformed is a flash flood's forte.

Andrea Booher/FEMA

Heavy rains caused severe river flooding in St. Maries, Idaho in 1995.

How They Happen

Almost anything in nature can cause a flood. That is one reason why flooding is the most prevalent and costly natural disaster. Nearly every community in the United States suffers some flooding problems. Most flooding is caused by the runoff from heavy storms. Intense rains saturate the ground and raise water levels in streams, lakes, ponds, rivers, and canals. The soil and vegetation may not be able to absorb the excess water which then runs into existing rivers or flows freely over the land.

Andrea Booher/FEMA

Flood waters in 1995 came almost up to the tops of the fence poles around this barn in St. Maries, Idaho.

Landslides and avalanches cause flooding by displacing the water in lakes and ponds or by damming rivers and streams. Large landslides near oceans can also incite tsunami. Hurricanes cause coastal flooding, river flooding, and flash flooding with storm surges and accompanying rains. Earthquakes cause flooding via tsunami primarily but also by upsetting rivers and lakes and damaging dams. Wildfires and drought contribute to flooding long-term by destroying water absorbing vegetation. Volcanoes often send melted snow and ice rushing down the mountain to deluge the valley below. Tornadoes can damage or destroy natural and man-made dams freeing reservoirs of water. Blizzards and heavy seasonal snowfalls melt in spring to raise water levels in rivers and streams above their banks.

Man can also contribute to flood conditions by damaging vegetation (deforestation, fires, construction, etc.) and by building unsafe dams or failing to properly maintain dams. Poor management of floodplains exacerbates the damages and costs of river floods.

El Niño

Bad floods, bad weather, bad allergies, bad moods, and even bad hair days are blamed on El Niño. The oceanic-atmospheric system is indeed culpable for unusual occurrences, especially relating to weather patterns. El Niño does not refer to the torrential downpours in the southwestern United States and west coast of South America or droughts and dry weather in Indonesia and Africa. These are consequences of the occurrence of El Niño.

Normally, trade winds blow from east to west across the South Pacific. These winds carry warm water from the surface of the ocean to the western coasts. Cool, nutrient rich water rises up in the East where small fish species thrive. About once every 4 to 7 years, the trade winds relax or even reverse direction. The warm water in the West sloshes back toward the East like water in a giant tub. Rainfall follows the warm water eastward causing severe flooding in South

America and leaving the western countries dry and prone to drought.

If this were all, it would only affect the tropical countries on both sides of the South Pacific. However, the rise in warm humid air that follows the ocean currents nudges the jet stream off course affecting global atmospheric circulation. Weather patterns then change all over the world.

The warm waters usually appear off the coast of Peru around Christmas, giving reason for the name. El Niño means "the child" in Spanish and refers to the Christ child when it is capitalized. The phenomenon is also called "El Niño/Southern Oscillation," or ENSO, due to the see-saw pressure systems in the South associated with the circumstance. El Niño normally lasts 14-22 months but, as in the case of the 1982-83 El Niño, can continue much longer.

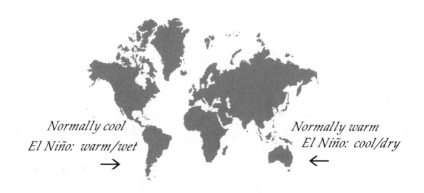

Normally cool
El Niño: warm/wet
→

Normally warm
El Niño: cool/dry
←

The Destruction

Floods are the most widespread and costly natural disasters in the United States. Floods have caused greater loss of life and disruption to families and communities in the United States than all other natural disasters combined! Approximately 90 percent of presidential declarations of emergency or disaster are due to events in which flooding is a major component. Floods cost the nation between two and four billion dollars (including agricultural production losses) and cause 200 to 300 deaths each year. Approximately 300,000 Americans are driven from their homes by floods annually.

Damage from flooding is diverse and can be extensive. Crops are drowned and washed away affecting supplies and prices elsewhere. Agricultural land is spoiled by loss of topsoil and deposition of minerals. Buildings, bridges, and other structures are damaged, eroded, or washed away. Transportation and communication are disrupted. Business and social systems are interrupted. Landmarks and local scenery are changed, marred, and dissolved. Water supplies and consumables are contaminated which lead to disease and health problems. Lives are disrupted with evacuations, loss of homes, and post-flood clean up and repairs. Wild animals, livestock, and pets are drowned along with man.

Andrea Booher/FEMA
The 1993 floods in the Midwest affected thousands of homes, like this one.

Andrea Booher/FEMA

Andrea Booher/FEMA

People's lives were interrupted and inconvenienced during the floods of the Midwest in 1993.

Andrea Booher/FEMA

Transportation is disrupted on this inundated road during the Midwest floods of 1993.

Andrea Booher/FEMA

Businesses across the country are adversely affected by flooding like this pub and McDonalds in the Midwest, 1993.

Andrea Booher/FEMA

This McDonalds kept its sense of humor with a marquee that reads, "waterfront dining".

What We Can Do

The best defense against flooding is to stay away from known floodplains. Strict regulation for development and habitation on floodplains would dramatically decrease the losses from floods. That is not very practical though, since floodplains offer some of the best living conditions, most fertile crop lands and choicest benefits of commerce. (This dilemma is one of the major problems facing floodplain management officials.)

The compromise is to set codes and standards to "flood-proof" buildings and homes. Structures are then safer and less vulnerable to rising waters and development is discouraged due to higher costs associated with meeting the standards. Developers are compelled to comply with regulations since federally funded insurance is only available to those who do.

The National Flood Insurance Program (NFIP) is administered by the U.S. government through the Federal Emergency Management Agency (FEMA). By adopting resolutions and ordinances to regulate structures in flood hazard areas, communities become eligible to participate in NFIP. Lower insurance premiums and other incentives are offered to communities who exceed the minimum requirements of the NFIP. FEMA works with communities to assess hazards, map flood zones, and accelerate floodplain management.

Flood damage can also be mitigated through the construction of flood control designs such as levees, dikes, channels, dams, catchment basins, and reservoirs. Historically, safety from floods was dependent on these types of projects. However, flood damage and losses continued to rise despite these water control efforts. It is, therefore, crucial to combine flood control with more prudent use of land and stringent development regulations.

If you are not sure if you live in a flood prone area, you can check with your local emergency management office, Red Cross chapter, U.S. Geological Survey, or Army Corps of Engineers. They can tell you the flood frequency and heights in your area, if your property is above or below the flood stage water level, and the history of flooding where you live. If there is a flood risk to your home, start preparing now.

Buy flood insurance through the NFIP since homeowners policies do not cover damages from flooding. Take pictures of your home and possessions and keep them with your important papers. Make sure your home meets the standards and regulations of your community. Make the necessary upgrades in order to comply with the codes. The investment will pay off in the long run.

Learn the alert signals in your neighborhood and the community flood evacuation plan. Your emergency management office or Red Cross chapter can help you get that information and tell you where shelters are located. Have a family plan (see Section 2) and include several evacuation routes if you live in a flash flood area. Teach all family members how to respond during and after a flood.

Prepare evacuation kits, emergency supplies, and a car kit (see Section 2). Keep your car at least half full of gas all the time. Have an extra supply of food and water including plenty of non-perishables (see Section 3). Keep your emergency supplies and food and water storage on upper floors where possible.

If a flood watch occurs, be prepared to evacuate. Move your patio furniture indoors and valuable possessions to upper stories if possible. Fill bathtubs, sinks, and other containers with clean water in case local water supplies are contaminated. Stay abreast of flood conditions and follow all instructions by authorities.

If flooding occurs, evacuate early. It is much easier to evacuate before water starts accumulating on roadways. Do not enter any flooded roadways. Only two feet of moving water will carry away most vehicles. If your car stalls in flood waters, abandon it and climb to higher ground. Do not attempt to move the car.

If you are on foot, move to higher ground. Do not enter water that is higher than knee deep. Swift moving water is very powerful and can knock you off your feet with only six inches or so. One cubic foot of water weighs nearly 63 pounds!

If you are indoors and unable to evacuate, go to an upper floor or roof if necessary. Bring bedding, clothing, and emergency supplies with you. If possible bring food and drinking water and wait for rescue workers to arrive. Do not enter the flood waters that are inside the house. There may be electrical hazards as well as contamination risks and poisonous snakes or animals.

When floods recede, do not return to your home until you are instructed to do so. When it is safe, inspect foundations for cracks and damage. Use caution when entering buildings. If the door sticks when you try to open it, it could mean that the ceiling is weak or about to fall. Stand outside of the doorway to avoid any falling debris. Wear good rubber-soled shoes and only use battery powered appliances. Be cautious of broken glass, sharp rocks, nails, and other hazards that are part of the post flood rubble.

Check for structural damage to windows, doors, walls, and ceilings. Inspect your home for fire hazards such as flooded electrical circuits, submerged electrical appliances, soaked gas control valves, or gas leaks. Do not use any electrical appliance that has been wet since there is a possibility of electric shock and/or fire. Do not use gasoline powered machines, generators, camping stoves, or charcoal indoors. Fumes and carbon monoxide exhaust from these items can be deadly.

Check your utilities for damage. If you suspect a gas leak, turn off the gas and call the gas company from an outside phone. Electrical damage is evident by broken wires, sparks, and/or the smell of burning insulation. Turn off the electricity at the fuse box or circuit breaker as long as you do not have to step in water to do it. (Call an electrician if water prevents you from shutting off the electricity when you suspect damage.) Sewage and water pipes may be damaged or backed up with contaminated water. Avoid using toilets or drinking water from the faucet if this is the case. Call a plumber to make repairs right away as damaged sewage systems impose a health hazard.

Start clean up and repairs to your home and belongings. (First take photographs for insurance purposes.) Throw away any food, cosmetics, and medicine that has come in contact with flood waters, including canned goods. Discard mattresses, bedding, carpet, and other materials that have been drenched. Rip out sheet rock, insulation, and wall coverings that have been soaked by flood waters before rebuilding. Flood waters contain sewage, chemicals, oil, and other health hazards picked up along the way. Silt and mud left after the water recedes should be shoveled then hosed away — both inside and out. Pump any water out of basements and cellars gradually to avoid structural damage. Think about moving out of the floodplain to avoid going through the same thing again!

Print Artist Stock

Thunderclouds roll overhead which could mean rain, hail, heavy winds, lightning and even tornadoes — but don't take shelter under the tree!

Severe Weather

"Behold, the Lord hath a mighty and strong one, which as a tempest of hail and a destroying storm, as a flood of mighty waters overflowing, shall cast down to the earth with the hand."
Isaiah 28:2

Weather is the mechanism that distributes the sun's energy. The sun heats the earth and stirs up the atmosphere causing winds and weather. The cold polar air circulates to cool the equator regions and the warm tropical air distributes heat to the poles. Water from the oceans evaporates into the air and moves over land where it condenses and wets the ground. One large thunderstorm can contain as much as half of a million tons of water! We depend on weather for our existence, our comfort, our familiar environment. When it is mild, we hardly notice that it is there. When it is severe, its strength and power awe us, terrify us, and sometimes catch us off guard.

Blizzards And Extreme Cold

The worst blizzard recorded in the United States killed 400 people. It happened on March 12, 1888, and it brought the Northeast to a halt. Communication was interrupted as all the telegraph wires were knocked down by the storm. Transportation was suspended as huge drifts of snow blocked roads and railways. Businesses ceased to operate as people could not get to work and could not function under the terrible conditions.

On Saturday, March 10, 1988, the New York newspapers had forecast cooling but fair weather for the weekend. By Sunday the 11th, it was raining from Washington, D.C., all the way north to New England. In New York, the largest city in the nation at that time, the wind picked up and temperatures dropped to freezing. Just past midnight, in the first few minutes of March 12, the rain changed over to snow. When the surprised New Yorkers got up in the morning, they were greeted by ten inches of snow and a strong wind.

Although many people tried to get out to work and other places, they soon realized it was futile. The trains stalled in the huge snow drifts and slick railways. The low visibility and icy tracks led to a collision near 76th street. One of the engineers died and twenty passengers were injured. Thousands of people were stranded in rail cars that could not move. One train traveled only two blocks in over six hours before trapping all of its passengers between stations. The trains were elevated so evacuation was difficult. The passengers eventually climbed down on ladders or waited for rescue. Other would-be travelers waited uselessly on train platforms in the bitter weather.

The roads were no better. Horses were doubled up to try to pull their wagons through the snow and wind. Even four or six horses could not overcome nature's impediments, so many abandoned wagons added to the hindrance. Ships in the harbor banged into one another and the piers in the winds gusting up to 90 miles an hour. Up and down the eastern seaboard, at least one hundred men died from drowning or exposure to the cold. One hundred and ninety-eight boats were sunk, damaged, or driven ashore.

By Tuesday, the snow began tapering off and people started digging out, which was not an easy task as the snow had hardened from the wind and cold. There was no place to put the tons of snow except in the middle of the street. In New York City, 21 inches of snow fell and in other parts of the Northeast it was even worse. Albany and Troy and other parts of upstate New York received around 50 inches of snow. Parts of Connecticut, New Hampshire and Vermont got 36 to 42 inches. Philadelphia only suffered ten inches but it fell on top of a layer of ice that came before the snow. From Maryland to Maine, nearly 25 percent of the United States' population was affected by the terrible blizzard of 1888. Exposure to the cold, accidents, and weather related illnesses took the lives of 400 people, half of them in New York City alone.

Ninety-five years later, on the same date, another winter storm hit the United States. March 12-15, 1993 saw a storm that affected half of the United States and would come to be known as "The Storm of the Century" or "The Superstorm of 1993." The low pressure system intensified in the Gulf of Mexico and traveled north into Florida and up the eastern side of America. Twenty-six states felt the storm's significance with record snowfalls and record low pressures reported in many places. It was the costliest blizzard in U.S. history.

Waves and storm surges battered the eastern coast from Florida to Maine. The hurricane force winds caused substantial damage and combined with a cold front to spawn over 50 tornadoes. Airports were closed from Georgia to Maine and road travel was icy and treacherous. Snow paralyzed cities in the Southeast that normally have little or mild snows. Over 250 people died and the total damages came to more than six billion dollars!

What They Are

Weather that is considered to be *extreme cold* varies with geographic location. In areas with mild winters, near freezing temperatures are considered extreme cold. In areas accustomed to fierce winters, below zero temperatures are extreme. When temperatures fall below about 20 degrees Fahrenheit and are accompanied by winds of 35 miles per hour or more with falling snow, conditions constitute a blizzard.

Winter storms are characterized by the region where they occur. Although all snow storms need cold, moisture, and rising air, the topography and climate affect the storm's features. In the Northeast, an area of low pressure off the mid-Atlantic coast moves northward causing havoc on the eastern states. Waves pound the shores from Virginia to Maine and coastal flooding ensues. Moisture from the Atlantic Ocean becomes snow and combines with the high winds to create blizzard conditions. Ice storms are also a problem since the mountains trap the cold air in place. As warm moist air moves over the region of cold air, it may rain. The rain falls through the cold air and freezes, falling as sleet or ice. This classic scenario is known as a Nor'easter and can be devastating to this densely populated region of the United States.

In the Southeast and all along the Gulf of Mexico, winters are normally mild. Once in a while, cold air moves far south through Texas and east across Louisiana, Alabama, Georgia, and Florida. When the temperatures drop below freezing and wet snow and ice develop, problems can abound. Automobile accidents increase because drivers are unaccustomed to icy roads and winter conditions. Some areas may not have plows, sand, salt, or other equipment to remove or treat snow on the roads. Many buildings are without heat or are poorly insulated against the cold. Roofs collapse and tree branches snap under the load of heavy wet snow and ice. Freezing temperatures can kill tender plants, including the citrus fruit crop that is grown in that area.

The Midwest and the plains states have severe winter weather as air masses from north, south, and west collide. Storms from Colorado move east or northeast across the plains and intensify from Canadian cold air and Gulf moist air. Canadian storms move south and southeast causing high winds and extreme cold. Temperatures in the northern United States experience wind chill temperatures as low as 70 degrees Fahrenheit below zero! (Wind chill is the perceived temperature when the effects of wind are combined with the actual temperature.) Cold air moving across the Great Lakes picks up moisture and creates small bands of snow known as *lake-effect snow*. Blizzards and extreme cold are common and debilitating in this middle band of the United States.

The west coast gets hit by storms that have moved across the Pacific Ocean and picked up moisture. Heavy snows fall on Washington and Oregon, Idaho, and sometimes California. When the snow hits the Rocky Mountains, avalanches occur, mountain passes are blocked, and freezing temperatures can be deadly. The wind squeezes through canyons and ravines picking up speed. The high winds (sometimes reaching hurricane speeds) combined with the snow can create severe blizzard conditions.

Alaska is known for its cold temperatures and brutal winters! Heavy snows accumulate in winter due to short days and little sun. Roofs collapse and trees are damaged under the weight of the snow. Glaciers form in the mountains, snow causes avalanches, and rivers get jammed with ice. The ice jams and the thawing snow can cause severe flooding. High winds combine with the extreme cold to create fierce blizzards and wind chill temperatures well below zero.

The National Weather Service issues advisories, watches, and warnings related to the winter weather conditions. A *winter weather advisory* indicates that conditions are expected to cause significant inconveniences, some of which may be hazardous (especially to motorists). A *winter storm watch* announces that severe winter conditions such as heavy snow and ice are likely within the next day or two. A *winter storm warning* means that severe winter conditions have begun or are about to begin in your area. A *blizzard warning* indicates that snow and high winds will combine to produce low visibility, deep drifts, and freezing wind chills. A *frost* or *freeze warning* means that temperatures are expected to drop below freezing which could cause significant damage to plants, fruit trees, and crops.

The Destruction

Extreme cold causes hypothermia and frostbite. Hypothermia occurs when the body temperature drops to 90 degrees Fahrenheit. Symptoms include: uncontrollable shivering, drowsiness, memory lapses, slow speech, and exhaustion. Frostbite is the damage to skin and tissue from freezing. It usually occurs to extremities and induces a loss of feeling and a white or pale appearance. The damage from frostbite can be permanent.

The cold weather also causes property damage and injuries indirectly. Blowing snow reduces visibility and combined with ice on roads causes car accidents and blocks roadways. Ice also jams rivers and storm waves batter the coasts. These winter by-products cause flooding and all the associated damages. Snow and ice cause structural damage as roofs collapse, trees fall on buildings and cars, and utility and phone wires snap under the weight. Deep drifts can halt transportation and trap people in cars, trains, buses, or even houses.

The heart works harder during cold weather to heat the body and the added exertion of shoveling, walking in deep snow, or pushing stuck cars can lead to heart attacks. River and road transportation is disrupted, crops and other vegetation die, oyster beds and fishing areas freeze. These disastrous disruptions combine to bring about food and fuel shortages.

What We Can Do

Every state except Hawaii is at risk for winter storms. Being prepared for winter ahead of the storms and cold is the best defense! By making sure you have all the supplies you need ahead of time, formulating a family plan, and winterizing your home, you can avoid crowds, sold out stores, inconveniences, last minute shopping, repair costs, and sometimes even injury or death.

Make sure you have emergency supplies, a first-aid kit, an evacuation kit, and a winter car kit (see Section 2). Augment your supplies with a snow shovel, ice scraper, rock salt, kitty litter (for traction), extra wood for the fireplace, fuel (if legal in your area), and extra blankets. Go over your family plan (Section 2) and make sure all your batteries are good and that supplies are easily accessible. If you use kerosene heaters, only use the correct fuel for your unit and only refuel outside. Store extra fuel if it is legal to do so in your community. Check with your fire department. Be sure to only use a kerosene fuel container for buying and storing kerosene. Never put your kerosene in a gasoline container, even if it is empty. Be cautious of placing your heater near flammables.

Now is the time to update your food and water supply as well (see Section 3). Include food that does not require refrigeration or cooking. Store only foods that your family eats frequently. When the weather is bad, it is very convenient to have extra food storage so you do not have to go out shopping for one or two items that you may need. If blizzard or storm conditions arise, you may have to eat what you have in the house as it is sometimes impossible to get to a store.

Winterize your home by weather-stripping windows and doors. Cover leaky windows with shrink plastic from the inside or storm shutters from the outside. Insulate walls, attics, and water pipes. When the temperature is expected to drop below freezing, leave faucets dripping a little to keep water inside the pipe from freezing. Repair leaky roofs and check stability as they may bear a heavy load of ice and snow. Keep storm drains and gutters clear to prevent further damage.

Winterize your car and check all the supplies in your car kit to make sure your batteries are good and everything works. Check or change the oil and add fresh antifreeze. Be sure the brakes, heater, and windshield wipers are all in good working condition. Fill the tires and, if necessary, add chains or switch to snow tires. Try to keep the gas tank full as much as possible. This prevents ice from getting in the fuel tank or lines and reduces your chances of having to walk in a storm!

Keep a watch on the weather. Get an NOAA (National Oceanic and Atmospheric Administration) weather radio from an electronic store (around 40 dollars) if you can. It will keep you updated on current conditions, watches, and warnings. Review storm terminology and the definitions of watches and warnings. Know what to do in case of flooding and teach all family members.

When cold weather hits, dress in layers that can be removed to prevent perspiring. Outer layers should be water repellent and mittens keep hands warmer than gloves. Stay indoors if possible and keep warm and dry. Avoid strenuous activity or over exertion when outdoors. Besides the added strain on the heart, bursts of activity cause sweating which then cause the body to chill in the cold weather. Do not drink caffeine or alcohol as they can accelerate the effects of the cold on the body.

If you are caught outdoors, seek shelter in a building or car. Cover your face to prevent frostbite and your mouth to protect your lungs. Do not eat snow or ice because it chills the body. Melt it first if you must use the snow to prevent dehydration. If you are in a car, flash your hazard lights and hang a brightly colored cloth from the antenna. Run the engine for around ten minutes each hour to keep warm. Remember to crack a window and clear the exhaust pipe to prevent carbon monoxide poisoning.

If hypothermia or frostbite is suspected, warm the victim's body slowly. Do not start with the extremities as the cold blood will circulate to the heart and could lead to heart failure. Make sure clothing is dry and use your own body heat if necessary. Seek medical attention as soon as possible.

Extreme Heat

Our warm blooded bodies regulate themselves to remain at the same temperature regardless of the surrounding environment. Since our bodies create heat, they must work harder to remain at 98.6°F (37°C) when the outside temperature gets hot. The hypothalamus at the top of the brainstem senses small increases in blood temperature and sends messages to the body to make corrections. The heart starts pumping more blood and vessels dilate to accommodate the increased blood flow. Capillaries near the surface of the skin open up creating a flushed look. As the blood nears the surface of the skin, small amounts of water evaporate as *insensible perspiration,* disappearing before they become visible on the surface.

If those changes are not enough to keep the temperature in check, millions of sweat glands on the outer layer of skin start working. They can shed large amounts of water and heat in the form of *sensible perspiration,* or sweat. Perspiration, both sensible and insensible, plays about 90 percent of the body's cooling down role.

When the blood circulates near the surface of the skin, it is only cooled if the outside temperature is less than the internal temperature. As the air heats to near body temperature, the blood cannot cool the body this way. Sweating becomes the major means for the body to rid itself of excess heat. The act of releasing moisture onto the surface of the skin does not in itself cool the body. The evaporation of the moisture is the cooling action. If the outside air is humid, evaporation is slowed. When outside temperatures are over 90 degrees (Fahrenheit) and relative humidity is more than about 75 percent, the body labors to keep itself within thermal limits. The skin flushes, the body sweats, the lungs pant to blow out hot air, but sometimes it is not enough. If the human body exceeds the upper allowable limit of internal temperature, it dies.

An average of 175 people die each year from extreme heat. Between 1937 and 1975 over 15,000 people died from the effects of heat and solar radiation. In 1995, more people died in the United States from heat related illnesses than all other natural disasters combined! In Chicago alone, over 450 people died from the heat that year.

When temperatures for a specific region are above the normal high by ten degrees or more for several weeks, it is considered extreme heat. If humidity is also a factor, it is generally referred to as a heat wave. When the effects of humidity are added to the actual temperature, the resulting perceived temperature is called the *heat index.* As the heat index increases, a *heat advisory* may be issued. This means that the heat will likely be an inconvenience for most people and a problem for some. An *excessive heat warning* is issued when the heat is expected to be dangerous for a large portion of the population in a certain region.

The Destruction

Conditions of extreme heat can be aggravated by dust storms and low visibility. Drought is often a factor because of little or no rainfall. There is also an increased demand for water during periods of heat. Food shortages may ensue due to the effects of heat and drought on agricultural production. Power shortages are an additional problem due to excessive use of electricity from air conditioners, evaporative coolers, and fans. In urban areas, pollutants are often trapped under the high atmospheric pressure which can lead to respiratory problems.

Overexertion, extreme heat, and high humidity can lead to sunburns, dehydration, and several heat related illnesses. Heat cramps are painful muscle spasms, mostly in the legs and abdominal area, due to heavy exertion. Heat cramps are believed to be caused from loss of water from heavy sweating. The treatment for heat cramps is cooling off, lightly stretching the painful muscle, and drinking a half-glass of water every fifteen minutes.

Heat exhaustion is a mild form of shock that occurs when the blood flow to vital organs is decreased. This happens because the blood is rerouted to the skin in attempt to cool it. Heavy work or exercise in hot weather heats up the body and brings on or augments this condition. It may cause weakness, fainting, and/or vomiting. If it is not treated, it will worsen and can lead to heatstroke. Treatment is applying cool wet cloths or towels, resting, and drinking a half-glass of water every fifteen minutes.

Heatstroke, also known as sunstroke, is extremely dangerous and even life threatening. At this point, the body's internal temperature control system quits working. The body quits sweating and internal temperatures rise to a point where brain damage and death can occur if not treated immediately. Victims of heatstroke may be unconscious, and could have body temperatures of 105 degrees Fahrenheit or more. Since it is so serious, call 9-1-1 immediately or other local emergency numbers and move the person to a cool place. Cool the body down quickly by immersing in a cool bath or by wrapping the body in a wet sheet and blowing it with a fan. At this point, it is better not to give fluids since the victim is likely to be vomiting and changing levels of consciousness.

What We Can Do

Since most disorders related to heat are due to overexposure or overexertion, the best defense is to keep cool and slow down. Stay indoors as much as possible and get plenty of rest. If you do not have an air conditioner or fan, stay on the lowest floor and close curtains over windows with direct sunlight.

Keep the heat out of your house by closing up spaces where hot air can enter. Weather-strip windows and doors and make sure spaces around air conditioners are insulated. Protect windows with shades or curtains as well as awnings or louvers. An awning or louver over the window can decrease the amount of heat entering the house by as much as 80 percent. On especially hot, sunny days, consider temporary reflectors such as aluminum foil on windows directly exposed to the sun.

Keep your body cool by wearing light weight and light colored clothing, as well as a wide brimmed hat for your face and head. Avoid sunburn as it slows the body's ability to cool itself. Drink plenty of liquids, even if you are not thirsty, and eat light meals. Alcoholic and caffeine drinks may seem refreshing but can actually enhance dehydration. Water is the best thing to drink on a hot day. In temperatures over 90 degrees, while in direct sunlight, your body can lose as much as a half-gallon of water every ten minutes! Do not take salt tablets unless directed to do so by your physician.

It is important to always have a good supply of food and water during periods of heat. Water and food shortages are common repercussions of heat and drought so you need to be ready. You will not want to use the oven or stove too much so it is good to have a selection of foods that you do not need to cook (see Section 3). Familiarize yourself with the symptoms of heat disorders and the treatments. Take a first-aid class if possible and update your first-aid kit (see Section 2).

Lightning

In April of 1942, a Shenandoah National Park ranger in Virginia was running from a fire tower during a thunderstorm. He was struck by lightning which went down his right side burning him all the way down his leg and knocking his toenail off. His name is Roy C. Sullivan and he still has scars on his right arm and leg to remind him of the incident. Twenty-seven years later, in July 1969, he was struck by lightning again! This time he lost his eyebrows. The very next year, July 1970, lightning found him for the third time burning his shoulder. Two years after that, in April 1972 he broke records by surviving his fourth lightning strike. (No other person has broken the "three strikes your out" rule by surviving more than three strikes of lightning.) His hair was set on fire this time. One year later, August 1973, he was knocked 10 feet from his car and again his hair caught fire from his fifth encounter with lightning. In June of 1976, lightning struck and injured him for the sixth time! One year later, in June 1977, he was fishing when lightning hit him for the seventh (and hopefully last) time, burning his chest and stomach. In Virginia he is known as the human lightning rod.

The death to injury ratio for lightning is higher than any other natural hazard at one to two. In other words, for every two people injured by lightning, one person dies. This high death rate for lightning victims makes it clear that survivors are very lucky. Perhaps the luckiest survivor of any person hit by lightning is Edwin Robinson of Falmouth, Maine. On June 4, 1980, he was out in the yard trying to bring his chicken in out of the rain. Edwin had been in a car accident nine years earlier that had left him blind and deaf. He learned Braille, wore a hearing aid, and used a cane to deal with his disabilities. While using his aluminum cane in a thunder-storm, he was struck by lightning and knocked face-down on the ground. He was unconscious for 20 minutes then awoke feeling very tired. Over the next few days, he re-gained his eyesight and hearing. The 62-year old man was cured. Not only that — his bald head started growing hair again!

How It Happens

Lightning is the discharge, or giant spark, of electricity across a potential that has built up within a cloud, from cloud to cloud, or from cloud to ground. Lightning can be pro-duced from blizzards, dust storms, and volcanic ash clouds but are mostly associated with thunderstorms. (See the chapter on Tornadoes for an explanation of how thunder-storms develop.) Particles within a thunderstorm cloud become positively and negatively charged most likely from collisions and molecular interactions. The violent air currents, as well as water and ice particles, that are rising and descending cause the charged particles to separate. The negative charges accumulate at the bottom of the cloud and the positive charges rise to the top. The net negative charge of the bottom of the cloud also induces a positive charge over the ground below.

The opposite charges have a natural attraction for one another and try to come together to neutralize the electrical potential between them. Since air is a poor conductor of electricity, a connection is hindered. The current can only cross the resistant air between by building a bridge of sorts. A small amount of invisible current, called a pilot leader, advances a short distance toward the ground. Another surge of current, called a step leader, moves down the first path and extends it a little farther. More step leaders continue to add a little length to the path moving it closer and closer to the ground. This sequence of steps repeats as many as 40 times until the path of ionized particles is near the ground. From the positive side (the ground), a discharge streamer or tracer extends upward toward the cloud. The discharge streamer is usually from a high point such as a pole, tree, hillside, antenna, or building.

When the step leaders reach down to one of the discharge streamers, the conductive channel is completed. A tremendous flow of electricity leaps upward along the completed pathway at about half the speed of light. (About 30,000 times faster than a bullet!) This surge of electricity is known as the return stroke or main stroke. The immense energy released causes the surrounding molecules in the air to glow. The light illuminates the downward pointing step leaders that were previously formed. This illumination of the step leaders is what we see as lightning. Although the main electrical discharge in a lightning flash is upward, our eyes perceive it as downward. The deception to our senses is partly due to the speed of the flash and also the downward pointing of the step leaders that are illuminated.

Once the return stroke has dissipated, dart leaders may advance downward to initiate three to five secondary return strokes along the same path. These continue until the electrical difference between the cloud and the ground is temporarily neutralized. The whole transaction from the origination of a pilot leader to the end of the secondary return strokes takes about one second.

The heat from a lightning stroke can reach 50,000 degrees Fahrenheit, or five times the temperature on the surface of the sun. The surrounding air expands suddenly from the heat which creates a sound wave. We know the sound wave as thunder. The flash of lightning and the sound of thunder essentially happen at the same time, but the light moves about a million times faster than the sound. In one second, the flash of light moves 186,000 miles but the sound only moves about 1000 feet or one fifth of a mile. (If you want to know how many miles away the lightning is, count the seconds between the flash and the sound and divide by five.) At close range, the thunder sounds like a detonating crack. Farther away, the high frequencies fade out and the lower frequencies bounce off buildings, hills, and other objects making a low rumbling sound.

The *fork* or *streak* lightning described above is the typical jagged bolt we see during storms. Sometimes the wind pushes the conduction path a few feet between return strokes and the lightning can look blurred as if somebody moved a camera during the shot. The resulting smear of lightning is known as *ribbon* lightning. When atmospheric particles interfere with the conduction path, the lightning appears notched or interrupted and is called *bead* or *chain* lightning.

Not all lightning occurs between a cloud and the ground. In fact, lightning within the cloud is five times more frequent. The flashes of light that illuminate the clouds with no discernible center are called *sheet* lightning. Sheet lightning can also occur from cloud to cloud in the same type of process. *Heat* lightning is flashes of light in the clouds that appear to have no associated sound. Heat lightning is really just lightning that is too far away for the sound to carry, about 15 to 20 miles or more.

The Destruction

Lightning strikes the earth 100 times a second. That adds up to 8,640,000 times every day! That represents around four billion kilowatts of continuous energy. At any given moment there are approximately 1800 thunderstorms happening around the world. With those big numbers, the potential for damage and death is wide.

There are between 100 and 200 deaths in the United States every year from lightning. That does not include those who die in fires started by lightning or those electrocuted by lightning-downed power lines. Lightning actually kills more people each year than hurricanes or tornadoes but avoids the media attention by taking its victims one by one. Most lightning victims are working or playing in open fields. The next most common group of victims are those who are boating, fishing, and swimming. Other dangerous activities commonly associated with lightning strikes are: working on heavy farm or road equipment, playing golf, talking on the phone, repairing or using electrical appliances, and taking shelter under tall trees. (Yes, people still do that!)

Lightning causes around 10,000 fires every year in the United States which results in the loss of two million acres of forest. (That averages out to more than 25 fires a day!) Not only forest fires, but house and apartment fires, as well as chemical and gasoline fires are started by lightning regularly. Damage to homes, property, and forests costs several hundred million dollars each year.

Animals are also killed by lightning and in greater numbers than humans. One reason for this is that animals huddle together in large groups so one shock from lightning jumps from animal to animal. In July 1918, a bolt of lightning in Wasatch National Forest in Utah killed a flock of 504 sheep! The United States Department of Agriculture estimates that lightning is responsible for over 80 percent of all accidental livestock deaths.

Thunderstorms also produce hail which can reach softball size in large storms. Small particles of water freeze, start to thaw, and re-freeze as they travel up and down with the currents inside summer thunderclouds. As more water particles attach to the hail, they grow larger and larger. When their weight is too great for the winds to carry them upward, they fall to the ground. Hail can fall at speeds greater than 100 miles per hour and causes around one billion dollars in damages to property and crops each year.

Straight line winds, also associated with thunderstorms, are identified to differentiate from the cyclonic winds of tornadoes. Strong gusts and sustained winds are very common with thunderstorms and are responsible for much of the storms' damages. The downward currents inside thunderstorm clouds are called *downbursts* and can cause damages equivalent to a strong tornado. Downbursts and updrafts are very dangerous to aviation.

There is no getting away from lightning and thunderstorms. Lightning visits every state but some states more than others. Florida leads the nation with 80 to 90 thunderstorm days a year. (A thunderstorm day is any day in which one or more thunderstorms occur.) The east coast has more lightning fatalities and the west coast has more lightning-ignited forest fires. Lightning can and does strike the same place more than once. The Empire State Building is hit by lightning an average of 23 times each year!

What We Can Do

The best thing you can do to avoid being struck by lightning is to stay indoors if thunderstorms are present or imminent. Since lightning is a giant spark of electricity, people associate it with common household electricity. It is a common mistake to think that wearing rubber soled shoes or the rubber tires on cars is adequate protection. Ordinary household electricity is 120 or 220 volts but lightning can be 100 million volts! After traveling miles through the relatively poor conducting material of air, a few inches of rubber will not stop it. It is so immense that it can cause non-conductive material, like wood and bricks, to explode.

An enclosed, hard roofed car will afford protection but not because of the rubber tires. The current travels the path of least resistance — through the metal frame and out the wheels. The current avoids the interior of the car since it would be a more arduous journey to the ground. When a person stands next to a tree, the lightning will often jump from the tree to the person on the way down as a body will conduct the current with less resistance than the wood.

Check the weather forecasts before planning outdoor trips and activities. Since weather watches and warnings are issued on a county or parish basis, know the name of the county or parish where you are and where you are planning to be. An NOAA (National Oceanic and Atmospheric Administration) weather radio will give you the latest updates and comes with an alarm that sounds if watches or warnings are issued. It can be purchased for about $40 at sporting, electronic, and hardware stores.

If you are caught outdoors when a thunderstorm approaches, take shelter immediately. If you hear thunder, you are close enough to the storm to be struck by lightning so do not hesitate to go inside. Get out of boats or swimming pools and stay away from water. Do not stand under a tree! (If you are in the forest, and there is no other shelter, stand under the shortest trees.) Stay away from tall structures like utility poles, towers, and fences and avoid natural lightning rods like golf clubs, fishing poles, bicycles, tractors, and camping equipment.

The best protection is inside sturdy buildings. Turn off electrical appliances and stay off the phone since lightning can travel through the wires. Stay out of the bathtub, shower, and away from sinks and drains because the metal pipes can conduct the electricity. If you cannot get to a building, get into a hard top car and roll up the windows. If there is no shelter nearby, squat low to the ground and rest on the balls of your feet to minimize contact with the ground. Do not lie down!

Thunderstorms and lightning often cause power outages so make sure you have emergency supplies and that all your batteries are good (see Section 2). Keep extra water and store food that does not require refrigeration or cooking (see Section 3).

Tornadoes

"And I looked, and, behold, a whirlwind came out of the north, a great cloud, and a fire infolding itself..." Ezekiel 1:4

The United States experiences about half of all the tornadoes in the world. Although they are most common in the central and southern Great Plains, no state is free from tornadoes. The worst tornado in New England history and most expensive single tornado (at that time) in the United States, happened in Worcester, Massachusetts. At 5:08 P.M. on June 9, 1953, a twister cut across the northern corner of the city. In less than one minute, it killed 94 people and sent 1,306 people to the hospital with injuries. The tornado caused 53 million dollars worth of damages to homes and businesses including a brand new six million dollar factory that had just opened a few days earlier.

One of the most hazardous elements of tornadoes is that they often occur in groups. When weather conditions are right, it is possible for dozens of tornadoes to be spawned on the same day. One such outbreak occurred on March 18, 1925. It proved to be quite deadly. A severe tornado and its offspring cut a 219-mile path of destruction from Missouri to Indiana, Illinois, Kentucky, and Tennessee. In less than four hours, the tornado, moving at 60 miles per hour, killed 689 people, injured over 13,000, and ran up 18 million dollars in damages to property. The episode, known as the Tri-State Tornado, was the deadliest in U.S. history.

Another deadly outbreak occurred on March 21-22, 1952. A group of 31 tornadoes killed 343 people in six states from Missouri to Alabama. The storms injured another 1,400 people and destroyed 3,500 homes. The property damages cost over 15 million dollars.

The largest outbreak of tornadoes occurred in an 18 hour period starting April 3, 1974. An unbelievable 148 tornadoes destroyed more than 600 square miles of land. The tornadoes, known as the Super Outbreak, killed 315 people and injured 5,500 more in thirteen states. Parts of ten states were declared federal disaster areas with the total damages topping 500 million dollars. More than 9,600 homes were destroyed and over 27,500 families suffered losses during the outbreak. Particularly hard hit was the town of Xenia, Ohio where 34 people died and half the city was leveled with almost 3,000 buildings demolished. The high school lost its roof and a school bus was deposited on the high school stage where a group of drama students had been practicing just a few minutes earlier.

April 8, 1998 saw a tornado, rated F-5, tear through Jefferson County in Alabama. The tornado packed winds close to 300 miles per hour and was a half of a mile across in some places. At least 34 people died including a young couple who left three young children with injuries and without parents. The storm injured another 250 people and left many homeless. Schools, churches, businesses and apartments were demolished along with 1,177 homes that were destroyed or damaged. Estimates of the damages reached 80 million dollars.

What They Are

A violently rotating column of air, often funnel shaped, extending from a thunderstorm cloud to the land is called a tornado, twister, or whirlwind. A tornado can be many different sizes, shapes, colors, and have a myriad of characteristics.

The ground path can be as narrow as a few yards to as wide as two miles. The average is between 100 and 250 yards wide and only a small percentage are larger than 400 yards wide. The length of the path of destruction can be a few feet or over 200 miles. A normal path is less than a mile but when the path is especially long, it is usually caused by several companion tornadoes. The longest continuous track made by a single tornado was over 300 miles.

Twisters usually travel across land at 30 to 40 miles per hour but can move as fast as 70 to 80 miles per hour. The duration can be as short as a few seconds to as long as several hours. Most tornadoes last fewer than 20 minutes. The wind speed varies from around 90 miles per hour to over 300 miles per hour. Some tornadoes are invisible except for the swirling dust at the base but most take on the color of the ground and debris that are sucked up into the funnel.

If a tornado occurs over the ocean, it is called a waterspout. The water is not sucked up from the ocean into the cloud as the name might suggest. The ocean water is swept up into a swirling cloud around the base of the tornado known as a spray ring. The funnel is white as it consists of condensed water molecules from the air and contains no dust or debris from land.

Large tornadoes can have several smaller ones surrounding it. These smaller tornadoes are called suction vortices and are often hidden in the dust cloud surrounding the base of the mother tornado. They spin with their own speed of rotation but they also rotate around the main tornado base and also move forward with it. Thus they have windspeeds much faster than the original tornado and are responsible for most of the damage. The suction vortices are small, usually between 10 to 100 feet, and only last a few minutes. Since they are usually hidden and are short-lived, evidence of their existence is normally found after they have passed.

Suction vortices were discovered and described by T. Theodore Fujita, the leading authority on tornadoes. A

Japanese scientist working as a meteorologist at the University of Chicago, he has been studying tornadoes for a half century. He devised the Fujita Tornado Intensity Scale to categorize the severity of tornadoes. Along with his colleague, Allen Pearson, he studied the effects of tornadoes and their damages and calculated the wind speeds necessary for specific results.

The scale was originally designed to have 12 rankings from F-0 to F-11, but wind speeds and intensities above that of an F-5 tornado have never been documented and are thought to be nonexistent. The scale therefore has six rankings. F-0 and F-1 are weak tornadoes and comprise 69 percent of all tornadoes. The strong tornadoes are ranked F-2 and F-3 and make up 29 percent of all tornadoes. The most violent tornadoes occur only two percent of the time and receive rankings of F-4 and F-5. Lamentably, this violent two percent of tornadoes accounts for 70 percent of all tornado fatalities.

Fujita Tornado Intensity Scale

Rating	Wind Speed	Damage
F - 0	40 - 72	Light
F - 1	73 - 112	Moderate
F - 2	113 - 156	Considerable
F - 3	157 - 206	Severe
F - 4	207 - 260	Devastating
F - 5	261 - 318	Incredible

Tornadoes occur in almost all parts of the world but are most frequent in the United States east of the Rocky Mountains. The most common time of year for tornadoes is spring and summer, but they can occur any time. Late afternoon into evening is a popular time for tornadoes to strike but they can occur at any hour of the day or night. In other words, tornadoes have a tendency to happen in the United States, east of the Rockies, in spring and summer, at dinnertime. However, they can and do occur anywhere, anytime of year, and at anytime of day.

What Causes Them

All tornadoes are triggered by thunderstorms, but not all thunderstorms produce tornadoes. There are three general conditions that must exist for tornadoes to occur. First, moist air must be unstable. Second, the unstable air must form elevated cumulonimbus storm clouds. Third, winds at high levels must blow at different speeds and directions from lower level winds. These conditions are all common in our weather and they often exist together. When the necessary elements coincide, it means that tornadoes are possible, not definite.

Stable air lies in layers and relatively little mixing occurs between the layers. Unstable air rises and falls and mixes with air of different temperatures and moisture contents. Anything that causes moist air to rise higher than its condensation level will make it unstable. For example, strong, uneven warming of moist ground will increase air temperature above the heated spots. The moisture will evaporate into the air and it will rise. As it gets higher, it cools, the water vapor condenses and the cooler air falls again. Depending on a number of atmospheric conditions, these updrafts and downdrafts can move at speeds of up to 100 miles per hour.

A mass of cold air moving into a mass of warm air can also instigate instability. The cold air pushes the warm air up because of its lower density. Just as material with less density than water will float to the top, air with less density rises above denser air. The warm air will continue to rise until it reaches air of the same density. As it cools, any moisture condenses into clouds and the cooler air falls. As more cool air moves in from the cold air mass, the warmer air is again pushed up.

Tornado Alley is a wide belt from Texas to Michigan where most of the tornadoes in the United States are produced. Atmospheric and climatic conditions in this area set up a perfect breeding ground for tornadoes. As cold air moves southeast from British Columbia, it meets warm dry air that is moving north from Mexico. The warm air rises above the cold air but has very little moisture so condensation is minimal.

As the air masses continue moving eastward, they meet up with tropical air moving north from the Gulf of Mexico. This air is very moist and warm. It is also pushed up by the cold air underneath and starts releasing heat energy in the form of condensation. Clouds form and rain and thunderstorms are common. The moist air is unable to release all of its heat energy as condensation because it becomes sandwiched between the cold air underneath and the warm dry air from Mexico above. As the warm dry air above starts to cool, the moist Gulf air starts to penetrate. This causes the rest of the heat energy to be released abruptly and clouds form as high as the stratosphere at 50,000 feet. Vertical air currents and intensified instability cause severe weather. Torrential downpours, lightning, thunder, wind gusts and hail are typical results. (See the chapter on *Severe Weather*.)

It just so happens that all of this happens pretty much under the jet stream. The jet stream is a ribbon of high altitude wind that blows in a relatively consistent pattern from west to east. When the cumulonimbus clouds reach such great heights, the jet stream can shear across the top, deflect the upcurrent and put a twist on the system. All three of the necessary requirements are then present and tornadoes are possible. This common scenario of colliding air masses over the mid-United States gives rise to the frequency of tornadoes there, thus, the nickname Tornado Alley.

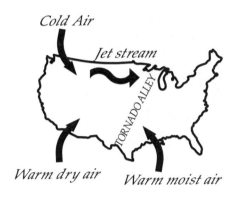

Tornadoes are also spawned from hurricanes that come onto the land. When this happens, the tornado is initiated in thunderstorms away from the eye of the hurricane. These tornadoes commonly occur along the Gulf of Mexico and all along the east coast.

There are many meteorological facets that can set up the necessary conditions for tornadoes to form. Weather is a complicated mixture of physics, thermodynamics, fluid mechanics, and other sciences. Almost anything can set up a "cause and effect" chain reaction that even the most sophisticated computers cannot sort. For this reason, the things that trigger tornadoes and promote and augment their likelihood, are not fully understood. Tornadoes are difficult to predict because we only know when conditions are suitable. The intensity cannot be predicted, only determined after the fact.

In addition, it is hard to study tornadoes because you have to be in the right place at the right time. Storm chasers go where the scene is set and hope that a tornado will make a dramatic appearance. Most experienced tornado hunters will spot only one or two twisters in a good year. That is, after driving hundreds or even thousands of miles.

Even when a scientist is lucky enough to catch a glimpse, the storm usually starts chasing the chaser. Video cameras and portable Doppler radar are used to make measurements and records but observations can only be made from a mile or two away. Some equipment such as TOTO (Totable Tornado Observatory) take measurements from inside a tornado but are limited by durability against strong forces.

Computer modeling helps us understand the behavior of tornadoes but relies on input of data. There is no possible way to feed the computer all the atmospheric, climatic, meteorological, and physical conditions surrounding tornado occurrences. Not only is it impossible to input the data, we do not even know (nor is it possible to measure) all the possible contributing elements.

The use of Doppler radar is improving tornado predictions by detecting rotational movement of water droplets inside clouds. Not all rotating clouds produce tornadoes so the information must be further analyzed. Doppler radar can only make measurements of storms that are less than about 100 miles away. The curvature of the earth and the lower resolution at greater distances interfere with accuracy. The direction of the storm's movement is also important because the Doppler effect is related to objects moving toward or away from the observer. (More than one Doppler radar from different vantage points overcomes this problem.) Although there are many limitations, with fortuitous placement and careful interpretation of results, earlier warnings are possible.

The Destruction

There are around 600 to 1000 tornadoes a year in the United States. The average is around 800 but some years have a higher incidence of twisters. The year 1973, dubbed "The Year of the Tornado," featured over 1300 tornadoes. There are approximately 80 deaths each year from tornadoes and another 1500 injuries.

The damage from tornadoes is mainly due to the force of high winds and the flying debris. The low pressure at the center, or eye, of the tornado is also known to contribute to the destruction. The amount of damage is dependent on the severity of the tornado and the area where it strikes. The higher intensity tornadoes obviously cause more damage and more fatalities. The more populated areas sustain more losses as do areas with mobile homes or poorly constructed houses. Most tornado fatalities happen to the elderly, people in mobile homes, people on upper floors or near windows, and people driving in cars.

The high winds of tornadoes not only destroy buildings, trees, cars, and other structures, but they also fling the debris around at extreme velocities. The missile-like rubble is responsible for most injuries and probably causes as much property damage as direct exposure to the wind. It only takes wind speeds of about 80 miles per hour to knock over mobile homes and tornadoes categorized as F-2 will pretty much destroy them. The more intense tornadoes with wind speeds over 200 miles per hour will destroy most structures and perform bizarre feats. There are stories of homes being lifted up, turned around, and set back down in the same spot, straws being driven into tree trunks, and cows, cars, roofs, and signs being carried miles away.

Another destructive force is the tornado's companion weather. Torrential rains cause flooding; lightning starts fires and strikes people; hail causes injuries and damage to property; gusty winds knock down power lines. Even if your home is not in the direct path of the tornado, the associated weather elements can be dangerous, uncomfortable, and at the very least, a nuisance.

Andrea Booher/FEMA

A tornado, rated F-4 hit the small town of Spencer, South Dakota on May 30, 1998. It wiped out all but six houses in the town of 350. Six people died and more than 150 others were injured.

Andrea Booher/FEMA

Spencer, South Dakota, 1998.

Andrea Booher/FEMA

Spencer, South Dakota, 1998.

What We Can Do

Early warning, advance preparation, and quick response are the keys to surviving tornadoes. Communities and neighborhoods should continue to improve warning systems, sirens, volunteer watch groups, public awareness, shelters, and preparedness education. Scientists and government groups should continue to study tornadoes and improve equipment used for monitoring and analyzing storm systems.

Make a family plan that includes periodic tornado drills. Get and update emergency supplies, evacuation kits, and car kits (see Section 2). Always have a good food and water supply (see Section 3). Take pictures of your home and belongings for insurance purposes and make sure that you have adequate coverage. Be prepared for severe weather, and educate your family about lightning, hail, and other dangers associated with tornadoes. (See chapter on *Severe Weather*.)

If tornadoes are prevalent where you live, find out about the warning systems in your area and any evacuation plans for the community. Know where the shelters are located and which one is closest to your home. Watch television reports and listen to radio updates when it is stormy out. It is good to have an NOAA (National Oceanic and Atmospheric Administration) weather radio with a warning alarm that will give you the latest reports from the National Weather Service. They can be purchased for around 40 dollars from many hardware and electronic stores. There is also a new generation weather alert radio called SAME (Specific Area Message Encoding). This radio will allow the programming to receive messages for a specific community as opposed to the larger area covered by a NOAA radio. It costs around 75 dollars.

If a tornado watch is issued, it means that tornadoes are possible. At this point, you should prepare to evacuate or move to a shelter. Locate all the members of your family and stay together if possible. Call neighbors and relatives living nearby in case they did not hear of the issued tornado watch. Pay close attention to storm updates and tracking and start getting your preassembled supplies. Occasionally, a tornado develops too quickly for warnings to be issued so pay close attention to developments outside. If you do spot a tornado, get to shelter or into the cellar immediately. If you have a chance, call 9-1-1 to report the tornado but always consider your safety first.

If a tornado warning is issued, it means that tornadoes have been spotted in your area, and you should take cover immediately. Tornado warnings are issued for very localized areas, so you probably only have a few minutes to get your family to a safe place. Since there is no way to predict the severity, you should prepare for the worst. The basement or storm cellar is the safest place to be in any building. The next best place is on the lowest floor in a room or closet with no outside walls. Stay away from windows and doors and get under a heavy table or mattress for protection.

It was previously believed that the low pressure at the center of the tornado passing over a house would cause it to explode. The difference in pressure between the inside of the house and outside could be equalized by opening a window on the downwind side of the house. Buildings do seem to explode during tornadoes, but it is now believed to be due to the high winds and flying debris rather than a sudden pressure difference. Opening a window only further exposes the house to the annihilative forces. In any case, most experts recommend forgetting about the window and getting to safety.

Anyone in a mobile home when a tornado warning is issued should leave to seek shelter elsewhere. This behavior is what some have called counterintuitive. Tornadoes occur during severe weather such as lightning, torrential rains, hail, and gusty high winds. It goes against instinct to leave the illusory refuge of a nice dry mobile home to the brutal weather outside. That is why it is so important to plan ahead. By discussing your options ahead of time and knowing the discomfort you may have to endure, you will not have to rely on instinct. (It also may encourage you to opt for more secure housing.)

If you live in an area where tornadoes are common, avoid buying or renting a mobile home. Homes without permanent foundations are not safe protection in any kind of natural disaster and are notably dangerous during tornadoes. About half of all tornado fatalities occur to people in mobile homes. There are more than 15 million people living in mobile homes in the United States and that number is climbing. Only one third of them live in mobile home parks where safety shelters are becoming more common. The other two thirds, or ten million people, live in rural properties where shelters are usually not within walking distance. That leaves a lot of people susceptible to the deadly energy of a tornado.

Andrea Booher/FEMA

Andrea Booher/FEMA

Mobile homes do not fare well during tornadoes. These homes from Gainsville, Georgia show typical damage from a tornado in 1998.

If you are driving, abandon your vehicle and go to the nearest building or low ground. If no building is near, lie face down, as flat as possible, and cover your head with your hands. Do not try to outrun the tornado or remain in the car. Make sure you are far enough from the car that it will not be blown over on top of you. Stay away from trees as the high winds may uproot them or break off large branches that could crush you. Lightning, which is prevalent where there are tornadoes, is another reason to avoid trees. Follow the same procedures if you are outside, away from shelter when you spot a tornado.

If you are at work, school, or in a public building when a tornado warning is given, follow the procedures instituted or the instructions given. If there are no emergency procedures or instructions given, go to an interior room, on the lowest level. Do not take shelter in a room with a high ceiling such as a gymnasium, cafeteria, or auditorium. Get under a heavy desk or table and turn your back to any windows that could not be avoided. Get into a crouched position and cover your face and head. Once the tornado passes, be on the lookout for more. Tornadoes often occur in batches and as long as weather conditions remain the same, tornadoes are possible.

Do not return to damaged homes until instructed to do so. Look for structural damage and check all utilities. (See chapter on *Earthquakes* for information on checking utilities.) Report problems to your insurance agent as well as pertinent specialists (electricians, gas companies, plumbers, etc.). Start making repairs and restock emergency supplies and food storage. Get ready for the next tornado!

Volcanoes

"And ye came near and stood under the mountain; and the mountain burned with fire unto the midst of heaven, with darkness, clouds, and thick darkness." Deuteronomy 4:11

The residents of Rodriguez Island, east of Madagascar in the Indian Ocean, thought they heard cannon fire on the morning of August 27, 1883. What they actually heard is considered to be the loudest noise in the history of man. It was the explosion of a volcano on the island of Krakatau, 3000 miles away! (That is like the residents of Los Angeles hearing an explosion that happened in New York City!) Sailors 25 miles away suffered ruptured eardrums and windows broke more than 200 miles away.

Krakatau

Rodriguez Island

Five cubic miles of rock, ash, and lava were ejected from the volcano with such a force that shock waves circled the globe seven times. Ash and cinders fell over an area of more than 300,000 square miles. Fine ash and dust particles were driven into the stratosphere as high as 50 miles. The cloud of dust and ash orbited the earth for two years. For a year afterwards, only 87 percent of sunlight reached the earth because of the dust in the atmosphere.

The island of Krakatau is located in Indonesia, between Java and Sumatra in the Sunda Strait. The massive eruption caused the volcano to collapsed on itself and two thirds of the island disappeared under water. Unfortunately, many villages and settlements on the island went down with the island. The seismic action of the eruption as well as the water rushing into the void caused massive tsunami. Waves 60 to 120 feet high pounded coasts of nearby islands and washed over the inhabitants of 295 villages. The catastrophe took the lives of more than 36,000 people.

Sumatra

Krakatau *Java*

Less than 20 years later, in another part of the world, another volcano erupted with calamitous results. Mount Pelée on the island of Martinique started rumbling and showing signs of life in April of 1902. By May 3, people were starting to perish. Halfway up the mountain, a fissure opened and spewed up boiling mud and steam. The fissure was directly under a village and 160 people died.

Two days later, a mudslide joined the River Blanche and a huge wave 120 feet high and a quarter-mile wide surged downward. It covered a sugar plantation killing 159 field workers. The mud wave continued down the river and dumped into the ocean. This created huge waves that pounded the downtown shores and killed hundreds more. In addition, the heat, mud, and poisonous gases on the mountain caused a plague of insects and snakes to move down into the towns.

Political strategies precluded any kind of evacuation even though there were plenty of signs of an impending eruption. The governor ordered troops to block all roads leading out of Saint-Pierre to prevent a mass exodus. On May 8, three days later, at 7:50 A.M. Mount Pelée exploded. A huge black cloud filled with lightning shot straight up into the sky turning day into night. A second cloud of super-heated steam, gas, dust, and ash rolled down the mountainside into the heart of Saint-Pierre. The glowing dust cloud moved at about 100 miles an hour and had internal temperatures of 1200 to 1800 degrees Fahrenheit.

The Caribbean

Martinique

In three minutes, nearly 30,000 people in the city of Saint-Pierre were annihilated. Some died from inhaling poisonous gases, some burned to death and others were literally boiled. The heat from the cloud was so intense that in many cases the body fluids were turned to steam immediately. Only two people in the city survived. One was a shoemaker who had barricaded himself in his basement. He was burned badly on the legs but was able to escape to a nearby village after the eruption.

The other survivor was a prisoner in an underground cell beneath the jail. The 19-year old, Auguste Ciparis, was convicted of murder and awaiting execution. Ironically, he was sentenced to hang on May 8, the day of the eruption. Instead of hanging, he was burned in his cell and was not rescued for almost four days. He recovered from his injuries, the charges against him were dropped, and he toured with Barnum and Bailey Circus until he died in 1929.

USGS

A view of the destroyed city of Saint-Pierre following the eruption of Mount Pelée on the island of Martinique in 1902.

Even with accurate predictions and effective evacuations, there is still a heavy price to pay for a major volcanic eruption. The first indication that Mount St. Helens was stirring was an earthquake nearby on March 20, 1980. Within a few days, hundreds of tremors were shaking things up. On March 27, the mountain blew a 250 foot crater out of its top and sent ash and smoke up into the air. Soon a second crater was created by another minor eruption. The craters soon became one big crater and after an eruption that lasted five and a half hours on April 8, it measured a third of a mile across and 850 feet deep. A bulge had developed on the north side and was growing at the impressive rate of five feet a day.

Authorities instituted a ban around the perimeter of the volcano called the red zone. The red zone was thought to be the biggest danger area should an eruption occur and nobody was allowed to enter this region. Journalists, campers, tourists, and the curious still entered this zone to get a closer look. People entered on old logging roads or went in at night or found other ways around the barricades. The erupting volcano was a spectacular sight and a cause for excitement. T-shirts were sold with several Mount St. Helens slogans. Washington was in the national news and this harmless display of nature was a once-in-a-lifetime scene.

Most of the residents that were evacuated obeyed the orders to leave, but a few remained. Harry Truman, a feisty 84-year old, owned a lodge near Spirit Lake and refused to go. He felt that the mountain was a part of him since he had been there over 50 years. He just knew that it would not erupt on him. Besides, he argued, if anything happened, they would have him off the mountain in five minutes. He was dubbed, "Volcano Harry," and enjoyed a few days of celebrity status as his stubborn resistance landed him several national interviews.

On Sunday, May 18, 1980, a 5.1 magnitude earthquake shook the ground directly under Mount St. Helens and caused three things to happen almost simultaneously. The bulge on the north side was jarred loose and started a massive landslide, the pressure underneath exploded laterally, and a vertical blast shot the top off the mountain. A geologist working for the U.S.G.S. (United States Geological Survey), David Johnston, was monitoring the mountain and reporting to headquarters in Vancouver. At 8:32 A.M. he reported, "Vancouver, Vancouver, this is it!" Those were his last words as he became an unfortunate victim of the volcano.

Harry Truman was never seen again as his lodge was buried beneath some 300 feet of debris. Loggers who were working nine miles outside the red zone were killed by the superheated cloud of gas and ash. Some of the victims were 18 miles away from the mountain when it erupted. There was no plan for search and rescue ahead of time. All the details were worked out on the spot. The first day, 130 people were rescued. The second day, around 25 people were found alive and saved. The third day, there were only about 15 to 20 lives saved. After the third day, there were no live rescues. In the end, at least 60 people were killed or missing.

The early warnings, evacuations, and the luck of a Sunday eruption saved countless lives and kept the death toll to a minimum. (Had the volcano erupted the next day, there would have been more than a thousand loggers working in the volcano's path of destruction.) The tally of devastation was nevertheless catastrophic.

When the bulge on the north slope was jarred loose, two thirds of a cubic mile of rock and ice rushed down the side of the mountain at about 160 miles per hour. When the debris hit Spirit Lake, it sent waves more than 400 feet out the opposite side. The displaced water and the debris laden bottom caused the lake to be 295 feet shallower than before. The landslide was driven by exploding steam and sent miles down the Toutle River. The heat had also melted 70 percent of the snow packs, which added to the volume. Sixteen million cubic yards of hot muddy water covered 14 miles. It filled the valley with a one mile wide boggy melange that was in places up to 600 feet deep. It was the largest landslide in recorded history.

Houses forty miles away were covered in mud from the landslide. At least 200 homes were destroyed or damaged, whole towns were evacuated, and hundreds were left homeless. Transportation and communications were disrupted. Seventy-five miles away, the Columbia River was choked with sediment which dropped water levels from forty feet to only nine feet. Around fifty ships were stranded or ran aground. Seventy million fish died in the Toutle River. Trees, bridges, buildings, wildlife, cut and stacked logs, machinery, cars, and other assets were swept away by the raging deluge.

The avalanche essentially took the pressure cap off the magma below. Shortly after the bulge slid down the mountain, the expanding steam exploded laterally with the force of a one-megaton bomb. The discharge sent rocks, ash, and blocks of ice in every direction for up to six miles. Strangely, the noise of the blast was heard in British Columbia, over 300 miles away, while people nearby did not hear a thing. (Scientists have not agreed on a reason for this odd occurrence.) The 500 degree Fahrenheit cloud of steam, ash, and poisonous gas shot northward and spread out at about the speed of sound. It quickly overtook the landslide and wiped out anything in its path.

An area 30,000 feet wide and 19 miles long was completely devastated. Trees were flattened or pulled up by the roots and thrown about like toothpicks. The logging industry appeared to be completely destroyed. All wildlife and vegetation over 230 square miles of land disappeared. Deer, elk, birds, and other animals died by the millions.

About the same time as the lateral blast, a vertical blast sent smoke, dust and ash 12 miles into the air. The ash-filled air blocked out the sun for towns as far away as 250 miles. By the end of the first day, the ash reached Montana. Within three days, it had followed the jet stream to the Atlantic Ocean in the East, dumping 500 million tons of ash over North America. In 15 days, the cloud had circled the globe to return to its place of origin.

The fall of ash buried nearby towns and caused massive transportation problems. On the ground, roads closed due to slippery ash and poor visibility. Cars stalled because the ash clogged air filters and damaged engines. Railways stopped running until tracks could be uncovered. Air travel also came to a halt as airports closed and air traffic was directed away from the ash cloud. Thousands of travelers were stranded.

Businesses were shut down temporarily and permanently as consequences of the disaster affected routines. Wheat, apples, hay, and other crops were covered with ash in Washington, Oregon, and Idaho. This resulted in crop damages of over 100 million dollars. Clean up costs were enormous and caused water shortages and sewage system problems in several states.

The economy and ecology of the area around Mount St. Helens were changed completely. The color of the land changed from lush green vegetation and sparkling blue waters to gray everything. The mountain itself lost nearly 2000 feet of its previous 8,364 foot height. The direct effects and far reaching, indirect effects are too numerous to inventory. The cost of damages and clean up is estimated at two billion dollars.

Washington

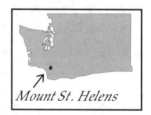

Mount St. Helens

USGS
Mount Saint Helens, 1980.

What They Are

Volcanoes are holes or vents in the earth's crust that allow the molten rock from the earth's interior to come to the surface. The molten rock is known as magma, and it forms reservoirs underneath the ground sometimes to a depth of several kilometers. When the magma comes to the surface, it can pour out as a liquid called lava or it can shoot into the air where it cools, hardens, and becomes fragments called pyroclastic material. Both lava and pyroclastic material add to the size of the volcano by building up around the vent.

Lava cools into rock after pouring out of the volcano. Depending on the consistency and chemical makeup of the lava, the cooled lava has several forms. When the lava cools in a rippled, rope-like fashion, it is call "Pahoehoe" (Pa-hoy-hoy). When it is jagged and looks like fragments of black glass, it is called "aa" (ah-ah). Legend says that aa is named for the sound made by the natives as they walk barefooted over the sharp rocks. Lava can also cool in big chunks called blocks or underwater where it is sometimes called pillow lava. When the froth or foam from the lava cools, it looks like a sponge and is known as pumice.

The pyroclastic material that shoots into the air also has many shapes and sizes. Big globs of magma with twisted or spherical shapes are called bombs. Bombs are larger than 64 millimeters and can be thrown thousands of feet. They usually have the appearance of a flat pie once they have landed. Rock fragments that were already a part of the volcano but are exploded into the air are called blocks. They have angular shapes and jagged edges. Small fragments that are either new, cooled lava or old parts of the volcano are called lapilli or cinders. They are usually about two millimeters in size. Ash is the smallest fragment of pyroclastic material and is blown with the wind. It can be carried into the stratosphere where it circles the globe. Ash is very damaging to airplanes as it clogs the engines and sticks to the windshield deleting visibility. It also blocks out some of the sunlight that reaches the earth which can cause long term weather changes. When pumice and ash consolidate to form a hard rock, it is known as tuff.

There are several types of volcanic eruptions. They are usually named for a volcano with that type of eruption. The least dangerous type of eruption is called Hawaiian. Like the volcanoes of the Hawaiian islands, this type has mainly lava flows and fountains and erupt with only a mild explosive force. This type of volcano is beautiful to witness and is probably the image most people get when they think of a volcano. Although very few deaths are caused by this type of volcano, they can be very destructive.

A Strombolian volcano is one that shoots glowing lapilli and bombs hundreds of feet high. Typically, gray clouds of ash rise thousands of feet into the air and lava pours from fractures at or near the base of the volcano. Strombolian volcanoes are known for their permanence and may continually sputter for months or even years.

A Vulcanian volcano blasts out hot blocks of old lava that have clogged the throat of the vent. It also explodes new bombs and lapilli from the magma below. Dark clouds of ash rise into the air forming giant cauliflower or pine tree shapes miles above. Lava flows typically come at the end of the eruption for this type.

Plinian volcanoes are extremely violent and send tall ash clouds into the atmosphere with explosions of atomic force. Generally, these types of volcanoes spew out a large amount of pumice.

The most violent and destructive of the volcanoes are those of the Peléan type. They produce glowing avalanches of super heated clouds. These glowing clouds of ash, pumice, and gases move up to 60 miles per hour (100 kilometers per hour) destroying everything in their paths. They hug the ground and behave almost like a liquid.

The type of eruption produced by a volcano generally depends on the amount of Silicon (SiO_2) in the magma. The less Silicon present, the more fluid the magma is, which allows the gas to escape relatively easy. This in turn produces a less explosive eruption as most of the gas has escaped without building up too much pressure. The more Silicon there is in the magma, the thicker and less fluid the magma is. Magma with a large amount of Silicon flows like stiff honey and does not allow gas to escape easily. This produces more explosive eruptions as there is more gas still present in the magma.

Magma with less than 55 percent Silicon is called, *basaltic*. Basaltic magmas usually have Hawaiian type eruptions. The volcanoes with basaltic magma usually are shaped with a broad gentle slope that looks like a shield. All the Hawaiian islands have a foundation of basalt and are shield volcanoes. At times, basaltic magmas produce Strombolian eruptions which cause a cinder cone to form. Cinder cones can be present on top of shield volcanoes. They are usually smaller than a third of a mile high and less than a mile in diameter and have a bowl shaped crater at the top.

When the magma has a silicon content of 56 percent to 65 percent, it is called *andesitic*. The eruptions of volcanoes with andesitic magma are usually not of the Hawaiian type. Volcanoes with this type of magma have steep sided tall cone-shaped forms. They are built of layers of lava and pyroclastic material and are known as composite volcanoes. They can be several miles high and up to 20 miles in diameter.

A silicon content of greater than 65 percent makes a magma called, *rhyolitic*. Rhyolitic magma is very viscous and does not allow gas to escape easily. When the lava oozes out of vents, it builds up into huge piles that stay within about a mile of the vent. If high pressure builds up in a volcano with rhyolitic magma, the explosion is usually of the plinian or peléan nature.

The character of the eruptions is not solely dependent on the amount of silicon in the magma. There are several other factors that determine the violence of the explosion. Another important element to consider is the amount of juvenile water present. When water changes from a liquid to a vapor, its volume increases more than a thousand times! The water present below volcanoes will stay in liquid form due to the huge amount of pressure underground. As the magma rises to the surface, the pressure decreases and the heat from the magma turns the water into steam. The vapor is confined inside the volcano and cannot expand. Eventually the pressure from the vapor overcomes the confining force of the volcano and it explodes. As water vapor makes up most of the gas that erupts, it is the main force behind the volcanic explosion.

How They Are Formed

The earth is made up of a dozen or so crustal plates that float on the molten mantle below. They push on each other and pull apart. Some plates move as much as a few inches a year. (See the discussion of plate tectonics in the *Earthquakes* chapter.) Most of the world's volcanoes are formed at the boundaries of the crustal plates. If the plates are pushing against one another, one plate will usually crumble and descend below the other (called subduction). This bottom plate will melt and become magma. The magma then works its way to the surface through one of the openings around the plate boundaries. If the plates are moving away from each other, the magma oozes up in the space left between them.

The large pacific plate is surrounded by several crustal plates that provide ample opportunity for volcanoes to form. The circular pattern surrounding the pacific plate is known as the "Ring of Fire" and contains 75 percent of the earth's active volcanoes.

There are several places, mostly underwater, where plates pull apart and magma can ooze up. These places form volcanoes underwater which can grow so large that they eventually form an island above sea level. Iceland is a good example of this type of volcano.

About five percent of the earth's volcanoes are formed from magma oozing up in the middle of crustal plates. These places are thought to be permanent "hot spots" that lie below the crust. The chain of Hawaiian islands was created by a hot spot underneath the Pacific crustal plate. As the plate moves across the hot spot, volcanoes form which grow from the ocean floor to an island above sea level. The chain of islands indicates the direction of the pacific plate's movement.

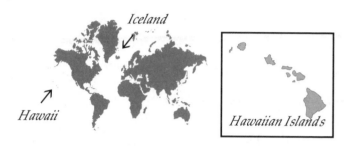

Iceland

Hawaii

Hawaiian Islands

How Eruptions Are Forecast

Each volcano is unique in its chemical and physical makeup and in its eruptions and characteristics. For this reason, active volcanoes are studied and scrutinized to better understand their "personalities" and eruption behaviors. Many of the world's active volcanoes have scientific observatories right on the side of the volcano. Hawaii's Kilauea is one such volcano and as the most active in the world, may also be the most studied. (Kilauea has been erupting since January 1983 and makes enough lava to fill 3000 railroad cars every day. It has produced enough new rock to pave a two lane highway that could circle the world 50 times!)

One precursor for all volcanic eruptions is earthquake activity. Virtually all eruptions are preceded by ground movement as the magma and gases underground go into motion. Sudden or intensified earthquake occurrence near a volcano can signal the beginning of disaster.

There are also other indicators that are common to most volcanoes and can be clues to a future explosion. A change in ground temperature may indicate that the hot magma is rising in the vent. The ground also may swell or change slightly as the pressure underground increases. A tiltmeter is used to measure these subtle changes. Gas composition at the site of fumaroles or openings in the ground can also indicate trouble. If only one of the indicators seems to be pointing to trouble, an eruption may not be successfully predicted or may not be imminent. However, if many or all the indicators are present, it is likely that an eruption will occur.

Volcanologists are very good at reading the signs and are continually getting better at predicting the volcanic outbursts. As they determine what is happening below the surface of the earth and they become familiar with the volcanoes' past activities, they can put together a reasonable estimate of the events to follow. Although a volcanic eruption can be sudden and unexpected, scientists can usually tell when a volcano is ready to burst and approximate what will happen. The severity of the eruption is harder to determine and has surprised us a few times!

The Destruction

As indicated by the stories at the beginning of this chapter, volcanic eruptions can be devastating and the consequences far reaching. The deadliest behavior is the pyroclastic flows that travel at high speed and are filled with extremely hot, poisonous gases. These superheated, glowing clouds of gas, cinders, and ash cannot be outrun and they demolish everything in their paths. Wildlife, vegetation, crops, homes, buildings, transportation systems, communication lines, and people are killed, burned, torn down, flattened, and consumed by this dreadful aspect of volcanic outbursts.

The ash that is spewed out of a volcano also causes severe problems. On the ground, it kills crops, blocks roads, stalls cars, weighs down ceilings, clogs rivers and streams, chokes people and animals, and turns everything gray or black. The clean up of fallen ash takes time, costs money, and uses up precious water supplies. In the air, ash blocks out the sun — turning day to night in nearby cities and changing weather patterns over a large area, sometimes for a year or more. Air travel is halted or re-routed to avoid the ash and travelers can find themselves stranded for days.

The heat that escapes the vent as steam or lava or both can cause avalanches and landslides that are devastating. (An avalanche caused by a volcanic eruption is called a lahar.) Often volcanic mountains are covered in snow or ice caps and huge amounts of water come pouring off the melted summits. The lahars can rush down the mountains so quickly that there is little time for evacuations and property salvage. If the water, mud, and debris are added to nearby lakes, rivers, or streams, the floods can wreak havoc for miles and miles. If the landslide hits a large body of water, tsunami will likely result which can devastate cities on the opposite shores.

The bombs, blocks, cinder, and rocks that are exploded from a volcano are deadly but only within a few miles of the eruption. Towns and villages built along the side or base of volcanic mountains are in danger of being buried by these projectiles as well as by the lahars. (The city of Pompeii was buried by pumice shot from Mount Vesuvius.) With the ability to predict eruptions ever improving, it is likely that such precarious towns would be evacuated prior to an explosion.

The lava that spews, pours, or oozes from a volcano causes fires and destroys property, wildlife, roadways, vegetation, and agriculture but rarely kills people. Lava moves slowly enough that evacuations can usually be handled successfully before the lava reaches the towns surrounding the volcano. Nevertheless, many people are left homeless and stand by helplessly watching their property and possessions destroyed by a force that cannot be stopped.

What We Can Do

There are over 600 active volcanoes. (Volcanoes are either active, dormant, or extinct. Not many scientists refer to volcanoes as extinct, however, since there have been eruptions from many volcanoes once thought to be extinct.) At any given time, there are between 12 and 24 volcanic eruptions occurring around the globe. Although some of these are underwater, many of them threaten people and property. One estimate states that by the year 2000, there will be 250 million people living in risk from these deadly explosions. Volcanic eruptions cannot be stopped. The only thing we can do is learn to understand volcanoes better, become increasingly accurate at forecasting eruptions, and evacuate the endangered. Therefore, the best way to protect yourself and your family is to listen to and heed all warnings.

If your area is evacuated due to the threat of an impending eruption, evacuate. It is especially important to have an evacuation kit if you live near a volcano (see Section 2). If roads are blocked off or closed, do not enter these areas. It may not be too difficult to go around the barriers or find back roads into the danger zones, but it would be very unwise. Scientists are getting better at predicting volcanic eruptions and are constantly monitoring the characteristics of active volcanoes. However, they still cannot predict the exact hour that the eruption will occur. You may be evacuated well ahead of an eruption or it may be hours or days. Take all warnings seriously and follow the instructions of your community leaders.

There are many dangers to life and safety when a volcano erupts. If you are within 100 miles of a volcano when it erupts, you will need to protect yourself and family. Depending on your distance and direction from the volcano, you will face different hazards. Depending on the individual characteristics and behaviors of that particular volcano, you will be at risk for different dangers. Find out all you can about the volcano in your area and about past eruptions. Find out if your town or neighborhood has an evacuation plan in case of an eruption. Does your children's school have a plan? Does your place of employment have a plan? Does your family have a plan? (See section 2.)

Even if you do not live near a volcano, you could be affected by an eruption. Food supplies could be damaged causing shortages and increased prices. Ash falls hundreds of miles away from the volcano could contaminate water supplies. There may also be water shortages due to clean up of ash falls. Weather in your area could be affected by a volcanic eruption far away. These weather changes could potentially last a year or more and affect food supplies. Every family should have extra food and water stored for emergencies (see section 3) regardless of the proximity of a volcano.

Wildland Fires

"...and the fire of the Lord burnt among them, and consumed them that were in the uttermost parts of the camp." Numbers 11:1

The summer of 1988 was the driest and windiest on record for Yellowstone National Park. It not only aggravated the flames of wildfires, it heated up an ongoing debate about fire management. Beginning in 1976, the park had had a natural fire policy that allowed naturally occurring fires to run their course without intervention. The plan included all the wilderness areas of the park, about 1.7 million acres, and had stipulations to protect visitor use areas and adjoining property governed by other agencies.

The reasoning behind the program was a desire to preserve or restore natural processes to national parks. Wildlife management oriented itself to minimizing changes in native environments by human influences. Since lightning caused fires are natural and have shaped the ecosystem for centuries, they should be allowed to burn.

The first fire of the 1988 season came on May 24 when lightning struck a tree in the Lamar Valley. It died out a few hours later when rains from the storm that started it, ended it. From the beginning of June, several other fires were started by lightning. Most were allowed to burn themselves out but a few were suppressed when they threatened other lands. By mid-July, the winds picked up, the rains disappeared, existing fires began to grow, and new fires cropped up in several locations. Officials began to worry and fear of the fires getting too big to stop, prompted more containment and suppression efforts. Despite protests from environmentalists, teams were called in to fight the fires.

Besides the natural fires, a large fire started by a woodcutter's discarded cigarette demanded the attention of the fire fighters. A team of the country's best fire behavior experts got together at the beginning of August to predict where the fires would be by the end of the month. The team of experts predicted that no more than 200,000 acres would burn. By August 12, the fires had burned 201,000 acres and the worst was yet to come!

By mid-August, there were at least 25 separate fires burning and winds were spreading them fast. More and more fire fighters came in to fight the conflagration. The flames did not come under control until September 11, when the first snow cooled the blazes and ended the fire season. The fires continued to smolder until November but they were no longer a threat to the park. By that time, over 1.5 million acres had burned in Yellowstone and surrounding forests. The use of 336 fire trucks, 57 helicopters and numerous retardant bombers were employed to control the burning. Forty-one bulldozers dug 850 miles of fire lines. At least 25,000 people worked on the fires and the total cost was more than 120 million dollars! In the end, the fires went out naturally.

Yellowstone National Park

The Debate

The Forest Service Administration, politicians, and especially the timber industry want fires suppressed and less use of controlled burns. Trees that could be logged and made into lumber for commercial use are lost to fire. Forests are destroyed and the beauty of green landscapes is marred by black burn areas. Wildlife is threatened and killed by fire and smoke. Towns are polluted as the smoke and ash from large fires drift over populated areas. People with asthma are affected by the poor air quality following neighboring fires. Homes and property are consumed by fires that should have been contained but got out of control. If natural fires are allowed to burn, the courts will be clogged with lawsuits from people who claim to be adversely affected.

Conservationists, environmentalists, other scientific associations, and nature alliance groups want naturally occurring fires in wild lands to burn freely. Forests have burned naturally long before man showed up to meddle with the intrinsic cycle of growth and adaptation. Where fires occur naturally, the ecosystem has adjusted and interference with fires has severe repercussions within the forest habitation. For example, the Douglas Fir has adapted by growing a thick bark on mature trees that makes them impervious to low intensity fires. They are also self-pruning which eliminates lower level branches and prevents fires from laddering their way to the top. When fires are frequently suppressed, a dense thicket of young trees grows and enables fires to get to the crowns of taller trees.

Aspen trees have a different method for survival of the species. When fire burns the standing tree, the root system is stimulated to send up thousands of stems or suckers. Scientists have counted as many as 60,000 suckers per acre following a fire. Since other trees that grow from seeds must first concentrate on building a root system, the Aspen has a good head start. As long as fires continue to occur, Aspens will dominate their areas. If natural fires are continually put out, there is no stimulant for the suckers to grow and Aspen trees age and weaken. There are also several grasses and shrubs that sprout from their roots. Fires rarely penetrate deeper than an inch or two into the soil so root systems usually remain intact. Fire does not destroy these plants, it only reduces dead surface material (fuel) and recycles precious nutrients back into the soil.

Lodgepole pines, which make up about 80 percent of the forest cover in Yellowstone Park, also have a unique way of dealing with fire. They produce open cones and serotinous cones. The serotinous cones are sealed with a resin coating that only allows the cone to open and release its seeds when exposed to the extreme heat of fire. In areas where fires are frequent, serotinous cones dominate and where fires are less frequent, open cones dominate. Obviously if fires are suppressed, the open cone will dominate and the tree will lose its unique ability to seed the area following a fire.

Fire's impact on wildlife is also less devastating than fire safety teachings and advertisements would imply. First, most animals can get to safety. Burrowing animals go underground where the heat does not penetrate. Birds nest in spring and their young have learned to fly by the time fire season starts in early summer. Other animals find refuge in the unburned areas of the mosaic pattern that fires produce. Second, the mosaic pattern of burned and unburned areas creates diversity and habitats with different qualities. Many species can find a suitable living environment within the many different options that are created by fires.

Studies have shown that species of plants, birds and mammals increase for 25 years following a fire. A clear illustration of this cycle starts with the round-headed wood bore. This beetle is attracted by smoke to recent burn areas and attacks standing dead trees. The large population of beetles attracts woodpeckers which start creating cavities in the trees. These snags become homes for many other species of nesting birds.

Without fires, a buildup of fuel in forests leave them vulnerable. A large fire that can get out of control and burn unnaturally strong is likely when suppression causes forest litter to pile up. The timber industry and congress have an answer for this. They suggest and actively campaign for thinning the forest through logging. That way, thousands of trees can be put to commercial use, employment in the lumber industry stays up and the forest will be thinned and less vulnerable to fire.

The environmentalists come back by pointing out the flaws in the logic of forest thinning. Thinning forests does not make them less vulnerable to fires. In fact, sometimes thinned areas burn more readily for a number of reasons. Usually the bigger trees have been logged and the saplings, which actually burn easier, are left behind. Also, thinning leaves wood chips, down and dead wood, and slash on the floors of forests which add fuel to the fire. Furthermore, thinned areas have more open spaces which means more oxygen to feed the fires that occur there.

To a person not familiar with the thinning process, it may seem to be a simple matter. It is not easy! It requires hundreds of men, endloaders, dump trucks, chainsaws and other equipment. After many days of long hours, only a few acres will be thinned properly. The costs of such an operation are staggering and could easily outweigh the commercial value of the lumber. It is physically and financially impossible to thin the fuels from the millions of acres of rugged lands.

The timber industry argues that fires cost millions in the loss of potential lumber. The environmentalists say that the cost of fighting the fires is more. In 1988, and again in 1994, the country spent over 500 million dollars fighting fires!

They also claim that the money is not justified since the fires are often not brought under control until it rains. The effects of nature are more influential to the outcome of the fire than the efforts of man.

Both sides agree that fires that threaten populated areas, visitor attractions, buildings and other structures should be contained. Fires that get out of control or have potential to spread and destroy private lands or property will be fought and suppressed. Those many fires started by arsonists or careless people will also be fought as they are not naturally occurring and upset the precious balances of nature. Reticulating humans and wildlands is delicate, dangerous, and brings opposing sides together to fight for protection.

What They Are

Fires that occur in wildland areas are a threat to people living near these areas as well as to those who enjoy recreational activities in the wild such as camping and hiking. Many of our national parks and their facilities are located within wildlands. Residential areas are expanding into relatively untouched lands and individuals are building away from crowded towns in remote regions. Wildland fires are a hazard for the entire human and wilderness interface.

Surface fires are the most common type of wildland fire. They burn along the floor of the forest and move rather slowly. Surface fires kill or damage trees and burn duff, litter and other fuels on the surface of the soil. *Ground* fires are usually started by lightning and, fittingly, burn on the ground. *Crown* fires burn along the tops of trees and spread quickly. Wind moves these fires rapidly and they are destructive because they can cover a lot of ground in a little time.

Wildland fires occur in every state and are classified in all wooded, brush and grassy areas. Four out of five forest fires are started by people — either by carelessness, or on purpose. Smoking in forests, not extinguishing campfires properly, lighting kerosene lamps near vegetation, and thoughtless or reckless use of flammables (such as fireworks) are common causes of fires. People also start fires knowingly to clear land or as a criminal act. When fires are started naturally, lightning is by far the most common instigator. Nature's fire starters also include the lava flows, pyroclastic material, and extreme heat from volcanic eruptions.

Photographer unknown

Firefighters work to save this house and the one on the following page that were overtaken by wildland fires in North Carolina.

Photographer unknown

The Destruction

There are an average of 130,000 wildland fires each year. Although naturally occurring fires (those started by lightning) have their beneficial place in the ecosystem, they are also a danger to man. Besides the danger of fire and smoke, heavy rains following a fire can invoke landslides and cause flooding. Erosion is also a problem where fires have destroyed the ground cover. Partially burned plants and trees are susceptible to disease and insect infestation which can then spread to other healthy plants.

What We Can Do

Living near wildlands requires special precautions. Following local building codes and weed abatement ordinances will help to protect your home. If you are planning to build or remodel near a wooded area, use fire-resistant materials. For example, use tile, stucco, metal siding, brick, concrete block, rock, and tempered safety glass. Do not use wooden shakes and shingles for the roof! Try to include a safety zone between your house and the forest vegetation such as a swimming pool or patio. Install electrical wiring underground if possible.

To find out if your home is at risk from wildfires, check with your fire Marshall, city engineer or planning and zoning administration. In general, if you live near wooded or grassy areas, you are at risk and need to check for and remove fire hazards. Prune your trees, removing branches lower than eight or ten feet and eliminating dead wood and moss. Mature trees should be at least ten feet away from your house but preferably farther away than the height of the tree. If you need to remove a tree, hire a professional contractor. It will probably cost between 300 and 500 dollars for a large tree. That may seem expensive but it is the same as the average home owner's insurance deductible. Rather than making a claim for losses later, prevent the problem early. Make sure your insurance policy covers damages from fires.

Clean your rain gutters regularly to avoid sticks, needles and other debris from piling up. Clean your chimneys every year. Keep the outside walls free of debris and combustible materials such as firewood and small shrubs. Clear your yard of shrubs, brush, litter and other debris for a distance of at least thirty feet surrounding your home. Keep your lawn mowed and rake leaves regularly, including underneath yard furniture and sheds.

Be prepared for the possibility of fires. Gather emergency supplies and make evacuation kits (see Section 2). Get a good hose and sprinkler to keep with your supplies. Teach each member of your family fire safety techniques and have a family plan that includes evacuation instructions (see Section 2). Have fire drills with your family on a regular basis and make sure every floor and all sleeping areas have smoke detectors. Check the batteries periodically and rotate any perishables in your kits. Have a good food and water supply and remember to rotate them regularly (see Section 3).

When fires are burning in your area, you can take a few more steps. Remove all patio furniture, umbrellas, tarps, and other combustible material from your yard. Close all your doors and windows and shut off gas valves and pilot lights. Remove flammable drapes, curtains, awnings, or other window coverings. Valuables that will not be damaged by water can be kept in a pool, pond or under a sprinkler. Keep the lights on to aid visibility in case smoke fills the house. If adequate water is available, leave sprinklers on the roof. Pay attention to local updates and be ready to evacuate if instructed to do so.

When the danger of fire has passed, check your house and property for sparks or embers. Check the roof and attic first and extinguish any potential dangers. Since hot spots can flare up for several hours, continue to check the roof, attic, and house for sparks or smoke. If your utilities have been affected by the fire, have a qualified person make repairs. See the *Earthquake* chapter for more details about checking utilities for damages. Clear your yard of burnt or partially burnt vegetation.

Section 2

Emergency Supplies

Family Emergency Planning

The most basic thing you can do to prepare for emergencies is to make a plan with your family. Even without necessary supplies, having a plan of action can help avoid fear, frustration, and anxiety. The plan itself will vary from family to family depending on the number and ages of children, elderly or handicapped members, locations of schools, work, etc. However, there are some general guidelines that will help you make your plan so it will be the most effective.

Preliminary Information

The first step is to contact your local Emergency Management Agency and American Red Cross chapter. They can tell you the types of disasters or emergencies that are most likely to occur in your area. The American Red Cross has material they can send you that will help you make your plans. Also, the local Emergency Management Agency can tell you what the warning signals are for your town or municipality. You will need to find out what the signals sound like and what they mean. They may also have suggestions of what you should do when you hear them.

Your sanitation department can tell you how you should dispose of garbage and human wastes should water and sanitation services be cut off. You need to find out what radio stations to listen to in case of emergencies, local weather conditions, school closings, etc. Locate the emergency shelters that are closest to where you are. It is a good idea to know of more than one, if possible, in case the one nearest to you is full. If you have pets, you may need to make separate arrangements for them since many emergency shelters do not allow animals.

In general, get as much information about your community's plans for emergencies or disasters as you can. Find out what your children's schools' and day-care centers' plans are as well as your employer's emergency plans. The information will help you as you start to create your own plan. It is also a good idea to check your home owner's or accident insurance policies to make sure you are adequately covered for the events that are likely to occur.

Creating Your Family Plan

Meet with your family to make your emergency plan together. If each person helps with suggestions and volunteers for specific responsibilities, they will be much more likely to remember what to do when the time comes. Share the information you have obtained about likely disasters, community plans, and sources of help. Talking about possible disasters can be frightening, especially to young children. When you meet to create your plan, make sure your family knows that you are trying to prepare them so that you can all be together and safe. Focus on the benefits of preparation and how it will help everyone to remain calm if disaster does strike.

As there are many different types of emergencies possible where you live, your plan will be general and should be adaptable to differing kinds of events. You will need to agree on a place to meet should an emergency or evacuation occur when you are not all together. Decide who will pick up the children and who will take care of any elderly, etc. You should all know the address of the meeting place and small children should carry the information (or give it to their schools, baby-sitters, day-care centers, etc.).

The American Red Cross suggests having an out-of-state contact since it is often easier to call long distance than locally after a disaster. All family members should call the contact to give their location and status. Teach your children to call 9-1-1 or other local emergency phone numbers if 9-1-1 is not available in your area.

Make sure you have smoke detectors on each level of your home and change the batteries at least once a year. This is an excellent responsibility to delegate to older children. Carbon monoxide monitors are also good to have on each level. Have at least one fire extinguisher in your home and teach everyone how to use it and where it is kept.

Somebody in your family needs to know how to turn off the water, gas, and electricity at the main switches in case you are instructed to do so. It is even better if all the members of your household know this information. Remember though, if you turn off the gas, somebody from your gas company may have to turn it on again.

Go through your house with your family to determine escape or evacuation routes from each room. If necessary, get window ladders to evacuate from higher levels. While you are going through your house, look for potential hazards. For example, loose book shelves or other large items that could fall during an earthquake, faulty or exposed electrical wiring, hazardous materials within children's reach, medicines or poisons not properly stored, flammable material near heating sources or pilot lights, and so forth. Also on your house tour, you should locate safe spots for earthquakes, floods, tornadoes, and other natural disasters.

Another stage of creating a family emergency plan is obtaining and organizing all the necessary supplies. Emergency supplies like flashlights, batteries, extra cash, and so on, need to be organized so you can find them without turning the house upside down when you need them. All your important documents should be together, preferably in a waterproof and fireproof box, and easily accessible.

You should have three-day survival kits for each member of your family which can be in backpacks or containers that are easy to grab and run. (Three days' worth of supplies is an absolute minimum and is mainly for evacuation purposes. See appropriate section for more details.)

Every family should have first-aid kits and car kits not only for emergencies but for everyday mishaps. Emergency supplies, first-aid kits, survival kits, car kits, food and water storage, and water purification techniques are discussed in detail in the following sections.

Part of your preparation is also for hard times such as unemployment, loss of money through accidents, law suits, medical bills, automobile and house repairs, etc. The best way to prepare for economic hard times is to keep an extra supply of food and water. This will also help when it is inconvenient to go to the store or when supplies to your area are affected by disasters in other areas.

It is always good to get out of debt and save money when it is feasible. By living within your means, you can avoid unnecessary interest that can really deplete any extra money you could have saved otherwise. Financial stability and planning are separate topics but are important to being prepared. You may want to consult a financial advisor or take a class to get all your money matters situated.

Other Suggestions

When you make your family plan, keep in mind that your family may have unique situations. You should not limit your preparations or emergency plans to only the things suggested in this guide. Use your own knowledge and skills to evaluate your family situation, house, town, etc. to build on these suggestions and tailor a perfect plan for you.

Go over your plan periodically to update it as situations change and to keep it fresh in your minds. The American Red Cross is a very helpful source that can help you answer questions about your area and where to locate emergency shelters or equipment. It is always good for as many people as possible in your family to take a first-aid class and a CPR course. These skills will help for minor emergencies and will help you to remain calm in the face of a crisis.

You may want to get together with your neighbors to create a plan to work together. This can be a valuable pool of skills and aid. It is also reassuring to have several families helping each other when it seems as if there are so many things to do and very little time. Check with your church or community groups to see if they have any suggestions, shelter locations, or emergency plans that will help your family. Although there are many things to do in order to perfect a family emergency plan, do not feel overwhelmed. Start today and improve your plan daily by accomplishing one task at a time.

First-Aid Kits

Every family should keep first-aid kits in both the house and the car. Since they are used for everyday mishaps as well as emergencies, they should be very accessible and not packed away with storage items. It is also important to know how to use a first-aid kit and all the supplies that you keep in it. You can and should take a first-aid course to better prepare yourself. Courses are often offered free or for a small fee from hospitals, community clubs, church groups, etc. Check your local paper for courses or call your American Red Cross chapter.

Below is a suggested list of items to keep in your first-aid kit. If you are starting from scratch, do not get discouraged if you can't find or afford to get everything at once. Each time you go shopping, get one or two items and gradually build up your kits. It is better to have a few bandages and tape than nothing at all!

- 12 Adhesive bandages of different sizes (including 2 extra large 2-inch bandages)
- 6 Butterfly bandages
- 1 Ace bandage
- 2 Sterile conforming gauze bandages (4x5)
- 2 Triangular bandages

- Rolled gauze
- Antibiotic ointment*
- Petroleum jelly or other lubricant
- Several types of tape including cling tape
- Non-adhesive dressings (3x4)
- 3 Sterile cotton tipped applicators
- Alcohol and/or alcohol preparation squares*
- Cleansing towelettes *
- Disposable instant ice compress
- 3-4 Tongue depressors (wooden)
- Thermometer
- Safety pins
- Tweezers
- Needle
- Scissors

- Pain reliever tablets (acetaminophen, ibuprofen, aspirin, etc.)*
- Antacid tablets*
- Anti-diarrhea medication*
- Laxative*
- Syrup of Ipecac*
- Sunscreen*
- Lip balm with UV protection*
- Sanitary napkins (can be used on heavily bleeding wounds)
- Latex gloves

*Watch for expiration dates and rotate every 12 months.

The quantities given are average numbers. If you have a large family or live alone, adjust the quantities accordingly.

Emergency Supplies

Below is a list of suggested items to keep in your home in case of power outages and other emergencies. Even if you already have the items, you should make sure they are accessible. One suggestion would be to have an emergency drawer, cupboard, or closet to keep the main items. Some things are better kept separate, such as first-aid kits and tools since they may be used for other purposes besides emergencies or disasters. Keep some items, such as flashlights, in several rooms or at least one on each level of the house. Evaluate your family's situation (ages of children, levels of the house, amount of space for storing emergency supplies, etc.), and use common sense to find the best places to keep your supplies.

Keep the supplies together that you will need to take with you in case of evacuation as part of your survival or car kits. This is a general list of emergency items and should not be treated as an all-inclusive supply. If you have special needs or other items that you feel you would need in an emergency, include them in your personal supply. Rotate batteries and perishables every six to twelve months or according to the expiration dates. Candles are not recommended for your emergency supply. According to U.S. Fire Administration statistics, up to four times as many people die in fires from candles after a disaster than from the disaster itself!

• Battery powered radio

• Flashlights

• Fresh batteries of appropriate sizes for radio, flashlights, etc.

- Matches
- Extra blankets and clothing (does not need to be kept with other supplies)
- First aid kit*
- Basic tools (screw drivers, hammer, wrench, pliers, etc.)
- Cash
- Can opener (non-electric)
- Electrical tape
- Paper cups, plates and plastic utensils
- Needle and thread
- Food* (especially ready-to-eat things such as canned tuna or fruit)
- Water*
- Chlorine bleach (regular, nonscented)
- Three day survival kits*
- Car kits*
- Extra toilet paper, soap and other personal hygiene items
- Plastic bucket or pail with tight fitting lid
- Extra garbage bags (plastic)
- Entertainment items (especially for young children)
- Fire extinguisher
- Special items (diapers, formula, medication, insulin, contact lens, denture supplies, etc.)
- Extra propane (if you have a propane burner, cooker, or gas grill)

*See appropriate section for more details

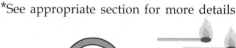

Car Kits

Keep your car kit in the car at all times. Include the following items and go through it periodically to update it, replace used items, change batteries, and rotate perishables. Even though the list is long, most of the items are small and the entire contents can fit in a relatively small space or in several places. During warm months, remove the items used for cold weather and replace them in the fall or winter. If you do not have a car, make sure all the items that would be in your car kit are included in your evacuation kits or emergency supplies.

- Battery powered radio (do not rely on car battery)
- Replacement batteries
- Flares
- Road maps
- First-aid kit
- Flashlight
- Spare fuse kit
- Fuse puller
- Jumper cables
- Blanket or sleeping bag
- Plastic poncho
- Tire sealant and inflator
- Standard tools (screwdrivers, lug wrench, tire iron, pliers)
- Shovel (pull-apart)
- Spare tire and jack

- Spare radiator hose
- Spare fan belt
- Work gloves
- Duct tape
- Bottled water
- Canned food items and opener
- Crackers or other snacks in glove compartment (for traffic jams, cranky children, etc.)
- Quart of engine oil
- Rope or chain
- Utility knife
- Bunge cord
- Fire extinguisher

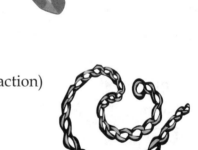

Cold Weather Items:

- Tire chains
- Sand or kitty litter (for traction)
- Snow scraper
- Matches
- Solar blanket
- 2 - 3 cans of "canned heat"
- Aerosol de-icer
- Candle

Warm Weather Items:

- Extra water

Evacuation Kits

(Also known as: Grab-And-Run Bags, 3-Day
Survival Kits, 72-Hour Kits, Etc.)

The purpose of this kit is to have all the essentials for you
and your family to survive for three days in case of an
evacuation. It is important to note that three day's worth of
supplies is a minimal amount. You should have no less than
one week's worth of food and supplies at your home. In cases
where outside help is demanded, the immediate response is
to the more populated areas, leaving many people in rural
areas to fend for themselves. In many instances, food, water
and supplies for three days will just not be adequate.
Therefore, it is best to rely on this kit solely for immediate
evacuation purposes.

The kit should be ready to "grab and run" at a moment's
notice. If you have a large family, it may be impractical to
keep all the items together in one container. Each member of
your family can have his own kit (such as a backpack) with
food, clothing, toiletries, etc. That way, the family kit (stored
in ice chest, duffel bag, garbage pail or other similar container) would be much smaller and easier to take with you.

Some of the items needed should already be in your car
kit or emergency supply. If they are already in your car, you
do not need to duplicate the items again here even though
they are listed. If some items are already in your emergency
supplies for the home, there are two things you can do. One
option is to keep the items in a container to throw in your
evacuation kit when necessary. The other option is to put
together your survival kit first, as the kit itself can then be
stored with your other emergency supplies to complete your

checklist. If you plan and store your emergency supplies and car kits wisely, you should not have to keep multiple supplies of the same item.

Remember to keep your keys to the house and car handy so you do not have to search at the last minute. If you have a hard time keeping track of keys it is a good idea to keep a spare set in your kit. Keep your important documents together so they can be taken with you when needed.

- Flashlight
- Radio (battery powered)
- Extra batteries
- Cash
- Blanket
- Nylon rope
- Matches
- Compass and map

- First-aid kit (or at least bandages and pain killers)
- Toothbrush and toothpaste (travel size is best)
- Toilet paper
- Sanitary napkins or other feminine hygiene products
- Deodorant, comb, razor and other toiletries
- Towel or dish cloth
- Change of clothing
- Whistle
- Trash bag
- Paper and pen/pencil
- Sewing kit (scissors, thread, needle, pins)
- Entertainment for children (books, crayons, etc.)

- Food and water for three days (see suggested list on following page)
- Can opener
- Knife, fork, spoon
- Any special needs (diapers, formula, medicine, etc.)

Three Day Menu

	Day 1	Day 2	Day 3
Breakfast:	Instant cereal	Instant cereal	Instant cereal
	Granola bar	Fruit roll	1/2 Trail mix
	Juice	Juice	Juice
Lunch:	Soup	Soup	Soup
	Crackers	Crackers	Crackers
	Beef jerky	Beef jerky	Beef jerky
Snack:	1/2 Trail mix	Granola bar	Fruit roll
	Fruit roll	Juice	Granola bar
Dinner:	Canned meat	Canned meat	Canned meat
	Canned fruit	Canned fruit	Canned fruit
	Granola bar	Pork & beans	Granola bar

Three-Day Food and Water Supply

The following is a suggested list of foods and liquids for survival and nutrition for three days. As the food and water are actually a part of your three-day evacuation supplies, package them together so they will be ready to grab and run if needed. If you know your family does not eat or care for a particular item(s), you should make substitutions according to your personal tastes. When you or someone in your family has special dietary requirements, you will need to make a separate list or menu that meets those needs. If you make your own list, make sure to only include foods that require no refrigeration or cooking.

Rotate the food kits at least every two years to maintain freshness, nutritional value and flavor. Check the expiration dates of all items that you put in your kit to be sure that it does not need a more frequent rotation schedule. Put a list of contents and the date on the outside of the kit.

Use creativity to make your food and water kits. It can be a family project to get all the items and package them in a container. Some individual kits may fit in a half-gallon milk carton which can be stapled shut and sealed with packing tape. (This particular list does not fit because there are too many canned items.) The milk carton can then be attached by duct tape to a two liter soda pop bottle filled with water. Once you get all the items together, you will be able to better judge the size of the container(s) you will need. Decide with your family if you want to make individual kits or buy foods in larger quantities and make one family kit. The quantities listed here are per person except where noted.

- 5 Fruit rolls
- 5 Granola bars
- 4 Juice boxes (4 oz)
- 3 Cans of meat (deviled ham, corned beef, chicken spread, tuna, vienna sausages, etc.)
- 1 1/2 Cups of trail mix
- 3 Packages of instant hot cereal
- 3 Packages of dry soup mix
- 1 Stack of soda crackers
- 3 Strips of beef jerky
- 1 Can of pork and beans (or one extra large can for entire family)
- 3 Snack size containers of fruit
- 2 Packages of chewing gum
- Assorted candy
- 2 Liters of water
- 1 Jar of peanut butter (for entire family)
- 1 Gallon equivalent of powdered milk (for entire family)
- 1 Box of graham crackers (for entire family)

Section 3

Food and Water Storage

Basic Food and Water Storage

How long could you and your family survive without going to the store? Any extra food and water you have can be considered a basic storage. However, without a conscious decision to store extra, and without some effort, this supply will fluctuate greatly and not be a reliable safety net.

The benefits of having an extra supply of food and water go far beyond disaster situations. Anytime you are low on cash, for whatever the reason, this extra supply will be a welcome blessing. Unexpected expenses, illness or disability, unemployment, wage decreases, and other economic setbacks can cause tight budget restraints. The ability to eat from your extra supply for a period of time may get you through the difficult times.

Anytime you cannot get to the store, for whatever the reason, you will not suffer for it. Bad weather, sick children, unexpected guests, car problems, work projects, time restraints, and other unavoidable situations can keep you from getting to the grocery store when you had planned. With a good food storage, you can put off shopping until a more convenient day or time.

The extra amount of food that each family needs is a matter of personal decision, space, eating habits, climatic conditions, and storage possibilities. Most emergency related groups recommend a two week supply of food and water. At least one source recommends a one year supply. I have included amounts for both two weeks and one year and ideally, you should be somewhere in the middle. The basic food storage suggested here is the amount of food that the average person would need to *survive* for two weeks and for one year's time. The expanded storage includes food necessary to supply total *nutritional needs* and allows for variety and tastes in your diet. The nonfood storage list contains necessities other than consumables that are recommended for your supply.

You may think that it would be very difficult to maintain a two-week supply, let alone build up a one-year supply of food because of money or space constraints. By buying a few extra sale items each week, you will be surprised how fast you can establish your supply. If you clip coupons and watch for sales on specific items, you can usually purchase a few extra things and still stay within your normal shopping budget.

By using a little creativity, you can store a large amount of food in amazingly little space. Take advantage of unused spaces in your house or apartment. For example, you can use the space under your bed, behind sofas, high closet shelves, etc. In addition, you can create spaces by making shelves in your basement, adding shelves to your closets or pantry, or making furniture out of the food containers. One example would be to stack large containers (like five gallon buckets), top with a piece of round wood and cover with a bright cloth to make a bedside or hall table. Be creative! Start with a two week supply and add to it as you can until you have the amount that is best for your family's needs.

Food stores best in dark, cool, dry places. Large fluctuations in temperature and/or humidity will cause quicker spoilage. For this reason, be cautious about storing food outside or in an outside shed or garage. You should not store food directly on a cement floor but wooden slats provide a solution for basement storage. Do not use glass containers if possible as breakage could pose a big problem.

Even though it would be easier to check dates and rotate foods, try not to store everything in one area of your house. If something should happen to that area (earthquake damage, fire, bug infestation, etc.) you would lose everything. Your food should be accessible (not in a vault in the attic for example) as you will want to use and replace foods in order to rotate the supply.

Although you can get good deals by buying in bulk, be careful about buying or storing large containers of single items. For example, if you have a huge sealed container of something and you need an extra cup one day, you would have to open the entire container and possibly waste what you cannot use.

Another important feature of food storage is the items that you choose to store. You need to have a variety of foods to fit your nutritional needs as well as your tastes. Use the list given here as a guide of the items and amounts needed for your two week to one year supply, but only store foods that you will use! This is very important. If you store containers of wheat as suggested but do not know how to cook with wheat and you do not have a wheat grinder to make flour, it is useless to you!

You can learn to use the items on the list that you do not normally use or you can substitute other foods in the same class (such as grains, vegetables, fruits) that you will use. It will also make it very hard for you to rotate your storage if you have items that your family does not care for or that you do not know how to prepare. It is easier to buy extra of the items that you are already using rather than making a special purchase just to comply with a pre-made list.

Use good judgment about the items you choose to store. Consider ease and knowledge of use, shelf life (follow manufacturers' instructions or "use by" dates), personal tastes of your family, ability to rotate, and ability to store according to recommendations. Make sure you have plenty of ready-to-eat items, such as canned fruits, tuna, nuts, etc., in case of power or gas outages that make it impossible to cook. Do not buy a large freezer and fill it with all your storage items as a loss of electricity will cause your whole stash to spoil.

One last hint that will help your food storage immensely: learn how to garden and store home grown foods! It is not only much more economical, home grown foods usually taste much better and it can be really fun to do. (Gardening is one of the most popular hobbies in the United States.) There are hundreds of books about gardening that can help you get started, teach you to garden in small spaces, show you how to keep the bugs off, and give you other details you might need. It can also be fun to can or dry your harvest, especially if you get together with your friends or neighbors and have a canning party!

Basic Food Storage

The foods listed are for survival needs and do not comprise every type of food necessary for total nutritional needs (see *Expanded Food Storage*). The quantities, where listed, are the amounts needed for two weeks and for one year respectively. Your food storage should be somewhere between these two values. Store a variety of products from each category.

Grains— 11.5 pounds per person (two weeks) to 300 pounds per person (one year)

Wheat, rice, oats, barley, enriched pastas, and other cereals.

Dairy Products— 11.5 quarts per person (two weeks) to 300 quarts per person (one year)

Powdered milk (Nonfat, dry, 4 quarts/pound), canned milk (evaporated), Parmalat milk, bottled or canned cheese, and other dairy products.

Sugar— 2.3 pounds per person (two weeks) to 60 pounds per person (one year)

Sugar, honey, jelly, jam, sweetened gelatin, sweetened drink mixes, and other products containing sugar.

Salt — 3 ounces per person (two weeks) to 5 pounds per person (one year)

Fats — 12 ounces per person (two weeks) to 20 pounds per person (one year)

Shortening (vacuum-packed), vegetable oil, butter, margarine, lard, etc.

Legumes — 2.3 pounds per person (two weeks) to 60 pounds per person (one year)

Dried soybeans, dried beans (pinto, navy, red, pink, lima, red and white kidney, small white, chickpeas, etc.) split peas or lentils, canned beans, canned nuts peanut butter, etc.

Garden Seeds — Enough for one planting season to fit your garden size

Multiple Vitamins — 14 tablets per person (two weeks) to 365 tablets per person (one year)

Yeast, baking soda, and baking powder — At least one package of each, more if you bake often

Expanded Food Storage

The foods listed here are to be added to the basic food supply in order to meet additional nutritional needs. The quantities, where listed, are the amounts needed for a two week supply and a one year supply respectively. Choose a variety of foods from each category.

Meats – 2.3 pounds per person (two weeks) to 60 pounds per person (one year)

Canned, smoked, dried, frozen or freeze-dried meats, poultry, and fish.

Fruits and Vegetables – 14 pounds per person (two weeks) to 365 pounds per person (one year)

Frozen fruits (peaches, strawberries, raspberries, cherries, etc.)

Canned fruits (peaches, pears, pineapples, fruit cocktail, tomatoes, etc.)

Dried fruits (raisins, prunes, apricots, bananas, apples, tomatoes, etc.)

Fruit juices (orange, pineapple, tomato, grapefruit, mixed berry, etc.)

Frozen vegetables (peas. corn, broccoli, mixed, cauliflower, squash, carrots. etc.)

Canned vegetables (green beans, peas, corn, mixed, cabbage, potatoes, yams, etc.)

Dried vegetables (potatoes, carrots, popcorn, etc.)

Vegetable juices (vegetable cocktail, carrot, beet, etc.)

Miscellaneous

Vinegar, spices, pickles, condiments, soups, special sauces (spaghetti, alfredo, gravy, etc.), corn meal, corn starch, etc.

Water Storage

Water may be the best place to start when it comes to storing extra. You can live longer without food than you can without water, and water is the easiest and cheapest item of your food storage!

Water — 14 gallons per person (2 week supply)

You will use seven gallons for drinking and seven gallons for other uses. (You need one gallon per person, per day.)

Store water in clean, tightly closed containers, away from sunlight. Water will store for six months to a year if it is clean when stored. Rotate your water supplies every six months to be safe. There are many commercially produced containers for storing water or you can improvise.

Empty plastic soft drink or juice bottles with screw tops are excellent for storing water because they are small enough to fit in extra storage spaces around the house or in the basement. These plastic containers are usually readily available around the house and it's a great way to recycle!

Empty milk jugs are not appropriate for storing water because the water will develop a bad taste. These types of containers also tend to leak after a while.

Water stored in bleach jugs becomes toxic for drinking but can be used for other purposes.

If the purity of your water is in question, you can treat it before using. In 1993, the U.S. Environmental Protection Agency, Food and Drug Administration, U.S. Department of Agriculture, Federal Emergency Management Agency, U.S. Public

Health Service, and the American Red Cross convened a water storage summit. They made several changes to the recommended procedures for storing water.

There are now only two recommended methods for treating your water:

1. Boil it vigorously for three to five minutes.

2. Treat it with liquid household chlorine bleach. Add 16 drops of regular, nonscented bleach to a gallon of water. Wait 30 minutes and then smell the water. If it smells like chlorine, it is acceptable to use. If it does not have a chlorine smell, add another 16 drops of chlorine bleach and wait another 30 minutes. If it still does not have a chlorine smell, discard it. If it smells like chlorine, it is acceptable to use.

The use of tincture of iodine is no longer recommended for treating water. There have been several studies showing that children under the age of eight, adults over the age of 55, and people with chronic liver, kidney, or thyroid conditions, can be seriously harmed by exposure to iodine in water.

For the same reasons, the use of commercially available water purification tablets is not recommended. Furthermore, an additional 30 million people in the United States are at greater health risk by exposure to halzone and similar products contained in these tablets.

If the water coming into your house is suitable for drinking, there is no need to delay starting a water storage. Start filling up containers today and before you know it you will have your 2 week supply!

Expanded Water Storage

You can store as much water as you want and have room for. The more you have in times of emergency the better off you are. You will also be in a position to help extended family, friends, and neighbors who may not be as prepared.

If you have a water bed or are thinking of buying one, check with the manufacturer to find out if there are special chemicals added to the water or the inside of the mattress. This will contaminate any water stored there for use as drinking water. Otherwise, it is a great place to have extra water. In any case, you can use that water for other purposes. Make sure you have alternative sleeping arrangements though!

Your water heater holds 30 to 100 gallons of water. If water in your area becomes contaminated, you must shut off the main water into the house so that the water in your heater will remain clean. Even with this precaution, you should treat any water from this source before use (see *Basic Water Storage*).

Large water or rain barrels are great because they can be kept outside. They come in different sizes and can be filled with a hose.

The water in the <u>back</u> of your toilet is clean. (Personally, I would treat it before drinking!)

The water in aquariums (fresh water) or fish bowls can be filtered and treated for drinking water.

If you are lucky enough to have a swimming pool, you have a great water storage! Use the treatment methods listed under *Basic Water Storage*.

If you have a water cooler and buy spring water from one of the many companies that deliver water, you might want to consider having the water delivered in a larger quantity. For example, you may want to purchase ten five-gallon bottles of water instead of the usual four or five. This depends on space of course but it is a great way to store water. The water is already drinkable, it is rotated regularly and the containers are supplied by the company. Some of the companies will allow you to start with ten and then deliver and pick up every three to five bottles that are used.

Do not forget that canned goods (especially canned vegetables) contain water. You should also have ice cubes in your freezer at all times. There are many ways to store extra water if you use your imagination!

Nonfood Storage

The items listed here are general suggestions for every family. Your family may have special needs which may require unique or extra storage. I have not included the amounts of most items that you would need for two weeks or one year. You need to gauge your family's use of products (such as toothpaste and toilet paper) and estimate the amount you would need for the amount of time that you want to store.

Bedding — Quilts, blankets, sheets, sleeping bags, etc. Enough to warm each person if there is no other source of heat available.

Clothing — Clothes and shoes for every season (including some in larger sizes to accommodate growth – and do not forget maternity clothes if pregnancy is a possibility), fabrics, patterns, needles, thread, yarn, etc.

Fuel — Coal, wood, paper logs, matches, battery powered light, propane (for outdoor gas grills), batteries, etc.

Paper Supplies — Toilet paper, facial tissue, paper plates and plastic utensils (to conserve wash water in times of shortage), garbage bags, aluminum foil, sanitary napkins, etc.

Cleaning Supplies — Soap (about one bar per month per person), shampoo, chlorine bleach (for water treatment), dish soap, toothpaste, household cleaning supplies, etc.

Financial Supplies — Extra money (savings account), traveler's checks, "cookie jar" cash (for times when you need money and the banks are closed for example), no debt (ideally), and emergency credit card.

Miscellaneous — Medicines, extra first aid items, anything you find yourself buying on a regular basis (i.e. diapers, wipes, pet food/supplies, vacuum bags, birth control products, contact lens solutions, etc.)

Hints and Suggestions

As you start storing food and water, you will discover what works best for you and a few things that do not work so well. Here are some suggestions regarding specific foods, storage, shopping, and preserving as well as some recipes, ideas, and resources that will help you with your plan for the future.

Wheat

Buy a hard variety of winter wheat (such as Red winter wheat). It should be grade #2 or better and have a moisture content of 10 percent or less. If it is not already packed for long-term storage, you will need to follow specific instructions for storing wheat. There are several books available that will tell you how to do this and you should be able to find one at your local library. You can also contact The Church of Jesus Christ of Latter Day Saints in your area as they store wheat regularly and will be able to put you in contact with somebody that can help you.

You should buy a wheat grinder if you plan to store wheat because flour is the easiest and best way to use your storage. You can also use a coffee grinder or food processor to crack the wheat which can then be used for sprouts or cereal. It is a good idea to learn how to use wheat and include it in your storage as it provides protein and vitamins and will store for long periods of time.

If your family is not accustomed to eating whole wheat, start slowly. Wheat is a natural fiber and will cause stomach problems that can be very uncomfortable if you suddenly switch to whole wheat. If you already have a lot of fiber in your diet, switching to whole wheat should be no problem. I am including some ways to use your wheat supply (other than grinding into flour) since cooking with whole wheat kernels is not as common as with other grains.

Wheat Cereal: Mix one cup of wheat with two cups of water and a half-teaspoon of salt. Put the mixture in a shallow pan or slow cooker. Bake it overnight at 200 degrees. If you prefer, you can soak the mixture overnight and cook it on the stove for about two hours. Serve it with milk and sugar (or raisins, dates, or other fruit). Grinding the wheat first in a food blender or wheat grinder will give cereal a finer texture.

Wheat Treats: Soak the wheat in cold water for 24 hours changing the water once or twice during this period. (Or boil the wheat for 30 minutes.) The wheat will triple in volume. Drain the wheat and rinse it. Remove the excess water by rolling the wheat on a cloth or paper towel. In a heavy kettle, heat some vegetable oil to 360 degrees. Put a small amount of wheat (about one and a half cups) in a wire basket or strainer and deep fry it in hot oil for 90 seconds. Drain it on absorbent paper. Season it with salt or other seasonings (as you would season popcorn), or boil one cup of honey and one tablespoon of water to the hard crack stage and pour over the wheat for sweet wheat treats!

Wheat sprouts: Place a handful of wheat kernels in a glass jar, cover the top with a piece of nylon net, and secure the net with a rubber band. Run water over the wheat, pour off the excess, and rotate the jar so the kernels cling to the sides. Place the jar on its side in a dark cupboard, then two to three times a day, dampen and rotate the jar. In about three days you will have healthy little sprouts. You can set it on your window for a day to green up, but this is not necessary. When the sprouts are about a half-inch long, just add kernels and all to your green salads, sandwiches, muffin batter, etc. You will find many ways to use them!

Wheat Chili: Brown together one pound of ground beef or ground turkey, one chopped onion, and one minced garlic clove. Add two cans of kidney beans (16 ounces each) or other beans, two cups of wheat (cracked with blender or grinder and cooked), four cups tomatoes and juice, one tablespoon chili powder, two tablespoons brown sugar, a dash of cumin (optional), and salt and pepper to taste. Simmer together for one hour.

Wheat Meat Balls: Beat one or two eggs and mix with two cups of cracked wheat (soaked 8 hours and well drained), one pound of ground beef, one bunch of chopped green onions, one teaspoon of beef seasoning, pepper to taste, and a half-cup of chopped walnuts. Use your hands to mix it thoroughly and form it into small balls squeezing to firm them. Drop balls into three to four cups of boiling beef broth. Cover, reduce heat, and simmer for one hour. Do not boil or balls will fall apart. Thicken broth after balls are removed and use as gravy. Worcestershire sauce and/or sour cream may be added to flavor gravy.

Rice

Buying and storing rice is simple. White rice, purchased in the grocery store will store right in the bags that they come in. I put several ten-pound bags inside a dark garbage bag and then put the garbage bag in a covered plastic container which I keep in a cool basement. Stored this way, white rice will store indefinitely! Brown rice has a shorter shelf life (about 6 months) because the bran layers contain oil which can go rancid.

It is easy to rotate the storage by keeping one bag of rice in the kitchen (I keep mine in a plastic jar with a lid) and when it is empty, use another bag from storage and replace the storage the next time you go to the store. Not only is it easy to buy and store, it is easy to use. Most people already know how to cook rice but if not, the instructions are right on the package for basic rice. You can use rice in breads, puddings, soups, salads, casseroles, desserts, and as cereal with warm milk and sugar.

Rice is very nutritious and readily available. (It is the principal food of almost half the human race!) Rice contains protein (high quality), all eight of the essential amino acids, thiamin, riboflavin, niacin, phosphorus, iron, and potassium. Brown rice also contains all the B complex vitamins and vitamin E. It has no fat, no cholesterol, no sodium, and no gluten. Make sure that when you buy white rice you get a brand that is enriched. If you want extra information on rice, you can contact the USA Rice Council at P.O. Box 740123, Houston, TX 77274. Telephone (713) 270-6699 or Fax (713) 270-9021.

Oats

Oats are delicious, versatile, and easy to store. I use only rolled oats since they are the easiest to find, buy, and cook. Another form of oats is whole kernels, called oat groats, which can be cooked just like rice, added to breads, or made into porridge. Also available are steel-cut oats (which are sliced oat groats), oat flour, and oat bran. Oat bran is believed to help lower blood cholesterol.

Follow the manufacturers' "use by" dates when storing oat products. Rolled oats last one to two years (call manufacturer if there is no expiration date) and can be stored in the original packaging. If bugs are a problem and/or flooding is a risk in your area, you can put it in something more durable than the cardboard containers it usually comes in. Plastic airtight containers will preserve it nicely. It is very easy to use as it can be eaten as cereal, in breads, muffins, cookies, etc.

Barley

The three main types of barley you can buy are hulled, pearled or flour. Hulled barley is the whole barley kernel with only the outer inedible layers removed. This type of barley is an excellent source of fiber and contains protein and the B vitamins. Pearled barley, the most commonly used type is not quite as nutritious since the fiber and nutrients are "pearled" away. Barley flour can be found in most health food stores and is a good substitute for wheat flour as a thickener. You can use it to make breads but it is best to mix it with other flours since it contains little gluten.

Storing barley is the same as rice. Barley can be used as a side dish alone or mixed with other foods or spices. Since it has a very mild flavor, it is best when cooked with stock or mixed with cream, butter, parsley or other flavor boosters. It is great in soups and casseroles. Be careful when cooking with pearled barley because it quadruples in volume! It would be very easy to turn soup into a soggy casserole. Hulled barley doubles in volume so you could use a little more of this type. I am including a great barley recipe, with permission, from a book called "Food Essentials: Grains and Pasta" by Carol Spier. You probably would not make this fancy dish during hard times but it is a great way to rotate your supply of barley in the meantime!

1 ounce dried mushrooms, such as shiitake or procini
1 cup hot water
2 tablespoons butter, more if needed
10 ounces fresh white or cremini mushrooms, cleaned
 and sliced
1 parsnip or carrot, chopped
3 shallots, peeled and chopped
1 onion, chopped
1 ½ cups hulled or ¾ cup pearled barley
2 cups water or broth, more if needed
2 tablespoons chopped fresh dill
Salt and freshly ground pepper (to taste)

Soak the dried mushrooms in the hot water for one hour. Drain through a coffee filter and reserve the liquid. Rinse the mushrooms in cold water and pick out any dirt. Dry and chop the mushrooms.

Melt the butter in a flameproof casserole dish. Add the chopped and fresh mushrooms and the parsnip or carrot, shallots, and onion. Sauté over medium heat, stirring frequently, until just brown, about 10 minutes. Add the barley and stir to coat. Add the reserved liquid from the dried mushrooms and the water or broth. Bring to a boil, cover, reduce the heat, and simmer until the barley is not quite tender, about 20 minutes. Stir in the dill and salt and pepper, adding more liquid if necessary, and continue to simmer until the barley is done, about 10 minutes more. (You can leave out the wild mushrooms if you prefer a milder flavor. Use all the liquids listed in the recipe even though you would not soak the mushrooms.)

Pastas

There are so many types of pasta and it stores well for about a year and a half so it is a great addition to your food storage. You will want to buy "enriched" pasta as it is more nutritious. Most families love to eat pasta so it is easy to rotate and easy to create a variety of dishes that everyone will eat. Make sure you store plenty of tomato sauce, spaghetti sauce, and other ingredients you like to eat with your pasta. (As good as pasta is, hardly anybody likes it plain!) Even butter and garlic on cooked pasta can be a delicious dish.

You can store it in the original packaging or remove the package and put it in a sealed plastic or glass containers. (There are containers that are tall and made especially for spaghetti, linguini, fettucini, and other long varieties of pasta.) The only problem with storing them in the original packaging, is you have to be careful with infestation. If your storage area becomes infested with bugs, they can get in the box type packages easily. You may want to put the original packages of pasta in a larger plastic bag and then place this in a plastic container with a lid (such as a hard plastic bucket).

You can also make your own pasta quite easily. It does not store well (about a week refrigerated or a month frozen) but it is fun to make and it tastes great. Just put about three cups of flour on a smooth surface and make a well in the center. Add four eggs in the well and beat slightly with a fork. (You can also add a little salt or olive oil if desired.) Little by little, add the flour into the egg mixture in the center. When the mixture is thick, but still wet, roll up your sleeves and dig in with your hands. Work in more flour and knead for five to ten minutes. If the dough sticks to your hands too much, add more flour as you knead. Cover the dough with plastic wrap and let it rest for 30 minutes to two hours. Roll small sections of the dough very thin by hand or with a pasta machine and cut into strips or other desired shapes.

If you want filled pasta, just place small amounts of filling in regular intervals along an uncut pasta sheet. Place a second sheet over the top and cut between the fillings. You should do this immediately after rolling out the pasta. When you are almost ready to eat, boil pasta in water until done. Although cooking time varies with different sizes, fresh pasta cooks much quicker than store bought, dried pasta. When the noodles rise to the top of the water, they are done. Be careful not to overcook fresh pasta as it gets tough. If you want to store the pasta for use at another time, dry it first by hanging on racks or sticks or by piling it loosely on a towel. After about four hours or so, place it in an airtight container and refrigerate it.

Powdered Milk

Always buy nonfat dry milk as opposed to dry whole milk. The lower the fat content, the longer it will store. The suggested shelf life for powdered milk differs greatly from one source to the next. I found literature suggesting a storage period from six months to three years for the same product! One manufacturer (Carnation) suggested storing their powdered milk for no more than one year from the date of purchase as the product performance changes. This does not mean that it will go bad in one year, but that it may separate, taste different, or have different properties than fresher powdered milk.

Since one year is about the average of all the storage times suggested, it is the time period that I suggest as well. (I have kept nonfat dry milk myself for a year and it was fine.) You need not worry, however, about drinking bad milk as you will know if it is rancid! If you can rotate your dry milk every few months, you will not need to worry. If you do not use it often, though, and your supply sits on the shelf for a while, you only need to smell it to know if it is good or not.

If your family does not like to drink dry milk (like mine), you can use your powdered milk supply when cooking. You cannot tell the difference in recipes like you can if you drink it straight. If you make your own bread, a tablespoon of powdered milk added to your dough makes the bread "velvety". You can also add a tablespoon or two to a glass of skim milk to make it richer.

Evaporated Milk

Add equal parts of water and evaporated milk to make "government standard" milk. Make sure you turn cans of evaporated and condensed milk about once every two or three months to prevent separating.

Cheeses

Dry cheeses, such as Romano or Parmigiano, will last for a long time when they are refrigerated. American cheese that is well wrapped will also last for several months in a refrigerator. You can freeze blocks of cheese or grate them first and then freeze. Do not store large amounts as a power failure will waste your supply.

Granulated and Powdered Sugar

One problem with storing granulated and powdered sugar is that it gets hard when it is exposed to humidity. Even though it is not as easy to use, it is still good. You can break it up by banging it with a mallet or tenderizing hammer, or you can put large chunks in a food processor or blender. You can also dissolve the sugar in water if it is to be used for sweetening drinks or caramelizing flan.

Another solution I have found, is double wrapping sugar bags inside plastic "food storage" bags. These can be purchased in the grocery store where you buy plastic and aluminum wraps. Put the sugar (in the original package) inside a storage bag, force out as much air as you can and secure the top with a twist or tie it in a knot. Then place this bag inside another bag and do the same thing. If you live in a very humid area, this will help but will probably not prevent hardening altogether. The best solution is to store sugar in air tight containers such as a bucket or canister with a rubber gasket in the lid. Sugar will store indefinitely.

Brown Sugar

Brown sugar gets hard when it gets dry. By covering it with an airtight, moistureproof bag, you can prevent hardening. It will store indefinitely at room temperature. If your brown sugar does harden, you can soften it several ways. You can put it in the oven on a piece of aluminum foil and heat it to 250 degrees until it is soft enough to work with it. (It will harden again as it cools.) You can also put it in a sealed container (such as a canning jar) with a damp paper towel, apple slice, bread slice, or other moisture source that is loosely wrapped in plastic so it is not in direct contact with the sugar. After about 12 hours, the sugar will absorb enough moisture to soften it considerably.

Honey

When honey is exposed to too much air and moisture, it crystallizes which hardens it. You can soften it up by placing the container in a pan of hot water. Follow manufacturers' "use by" dates. If none are given, expect to store honey about one year in a cool, dry place.

Salt

Unless you live in a very humid area, your salt should stay free of lumps and store indefinitely. If you find that your salt is starting to harden or lump, add a few grains of rice to the container and that should solve the problem.

Legumes

Dried beans, split peas, and lentils are cheaper, last longer, and yield more per pound than canned. For this reason, your supply should be mainly dry with some canned beans for convenience. Canned beans are good to have because if there is a power outage, they can be eaten cold out of the can. (Granted they are better heated!) The 60 pounds per person suggested amount is based on a supply that is mainly dried beans. If your supply includes a large amount of canned beans, you should store more than 60 pounds per person. (One pound of canned beans yields about two cups whereas a pound of dried beans can yield more than quadruple that amount!)

Pests

The best way to prevent pests from invading your food supply is to keep everything clean. Even the outside of containers should be wiped clean. Small amounts of flour or cracker crumbs on the outside of packages will lead weevils or silverfish from one package to the next. Keeping spilled crumbs and other food off counters and out of cupboards will go a long way in preventing bugs or the spreading of bugs.

A stick of spearmint gum inside open containers of flour, sugar, salt, other baking products, and wheat grinders will discourage uninvited intruders. If you do find that your open containers have been contaminated, the best bet is to discard them and any other nearby open containers of food. (Alternatively, freezing wheat or flour for several days will kill all the weevils and their eggs.) After discarding the tainted food, a good soapy scrub of the area should eliminate any problems.

You can spray insecticide in any areas that do not contain open containers of food or dishware. Cockroaches and ants can usually be controlled with bait traps. If you still have a problem, consider calling an exterminator.

If you have problems with mice, you can usually eliminate them using traps baited with peanut butter, chocolate, cheese, or other smelly foods. (There are traps that do not kill the mice but if you do not let them loose far away from your home, they will only return the same way they got in the first time!) In the meantime, keep your foods in canisters or lidded buckets as opposed to cardboard or thin plastic which they will chew right threw. Plug up any obvious holes or spaces underneath doors, near windows, in the basement, or any place you can see where they might enter your home.

If you are plagued by rats, you may have a harder time catching them in traps as they are very wily. First, you will need much bigger traps made specifically for rats. Second, when you see a rat or signs of rats, you probably have more than a couple of rats. (They have several large litters a year.) Once a few rats have been caught in the traps, the rest of the group will stay away from the traps no matter how enticing the food is.

Rat poison works well with rats but is very dangerous to have around if you have children or pets. The rats will also stay away from the poison after they start dying but usually they will all have eaten from the poison baits by the time they start to die. (It takes three to seven days to kill them but even one feeding will usually do the trick.)

Another problem with trying to poison them yourself, is you do not know where they will die. If they die out in the open, you have the messy job of disposing of the carcass. Worse, if they die in hiding, which could be inside walls or under floorboards, you have to find them or they will rot and stink. Since mice, rats, and other rodents all carry many serious diseases and cause structural damage to your home, it is a good idea to call an exterminator at the first sign of trouble with these types of pests. Do not mess around and let them multiply!

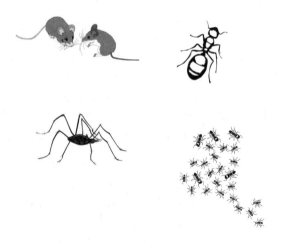

Afterword

There are a lot of suggestions in this book and you may feel overwhelmed at the chore ahead. Please don't. You do not have to do everything suggested in this book in order to be prepared for natural disasters. Do what you can. Start with a first-aid kit and a family plan and add on to your preparation little by little. I wish you all luck and I hope that your lives are never affected by a natural disaster. If you ever do experience the forces of nature, I pray that you will be safe and that your knowledge and preparation will help to protect and comfort you and your family. God bless you.

I would love to hear from you and encourage you to send me your questions, comments, corrections, criticisms or suggestions for future editions.

Kellye A. Junchaya
c/o MedCap
P.O. Box 2085
Clifton, NJ 07015

medcap.pub@juno.com

About The Author

Kellye Junchaya is a Bio-Engineer from Arizona State University. After getting her degree, she moved to a small town in Texas on the shores of the Gulf of Mexico. Within a month, hurricane Gilbert moved into the Gulf, compelling her to evacuate her home. Although the hurricane did not damage her city, it impacted her life. She has been studying natural disasters with fascination since that time. Kellye now lives in New Jersey with her husband and three children. She enjoys sports, camping, reading and especially being a mom.

Appendix
Additional Resources: Organizations, Academic Institutions, Government Agencies and Periodicals

**NOTE: Many of the addresses and phone numbers listed below are courtesy of the *Natural Hazards Observer*, special edition, 1997. The information is current through June of that year. If you wish to contact the agencies listed, it may be wise to look at the website or attempt an e-mail first as changes in personnel and/or subscription rates may have occurred since that time. For more information about the periodical, *Natural Hazards Observer*, see their listing below under "All Natural Disasters." This list is not all-inclusive.

All Natural Disasters

The American Red Cross
National Headquarters, Disaster Services Department
Armone T. Mascelli, Acting Vice President
8111 Gatehouse Road, Second Floor, Falls Church, VA 22042
Tel: (703) 206-7460
Fax: (703) 206-8835
E-mail: infor@usa.redcross.org
Website: http://www.redcross.org
***Disaster information is provided by *local* Red Cross chapters. Requests sent to the national headquarters are referred to local chapters.

Asian Disaster Preparedness Center
Asian Institute Of Technology, P.O. Box 4, Klong Luang,
Pathumthani, 12120, Thailand.
Tel: 66-2-524-5353
Fax: 66-2-524-5360
E-mail: adpc@ait.ac.th
Website: http://www.adpc.ait.ac.th/Default.html

Asian Disaster Preparedness News
Quarterly periodical–contact the ADPC information officer for subscription information
Same address, e-mail and website as listed above.
Tel: (66-2) 524-5378 or 524-5354
Fax: (66-2) 524-5630

California State University
Center For Hazards Research
Christine M. Rodrigue
Department of Geography and Planning, Chico, CA   95929-0425
Tel: (916) 898-4953 or 898-5285
Fax: (916) 898-6781
E-mail: crodrigue@oavzx.csuchico.edu

CCEP News
Irregular periodical–$15/year
Canadian Center for Emergency Preparedness, P.O. Box 2911, Hamilton, Ontario, Canada L8N 3R5
Tel: (905) 546-3911 or (800) 965-4608
Fax: (905) 546-2340
E-mail: ccep@netaccess.on.ca
Website: http://www.netaccess.on.ca/~ccep/ccep

Center For The Study of Emergency Management
1241 Johnson Avenue, Department 160, San Luis Obispo, CA   93401
Tel: (805) 782-6787
Fax: (802) 782-6730
E-mail: wbalda@simeon.org
Website: http://www.simeon.org/msm.html

Colorado State University
Hazards Assessment Laboratory
Hal Cochrane, Director
Fort Collins, CO   80523
Tel: (970) 491-6493
Fax: (970) 491-2925
E-mail: hcochrane@vines.colostate.edu

Disaster Emergency Response Association International
P.O. Box 37324, Milwaukee, WI   53237-0324
Tel: (970) 532-3362
Fax: (970) 532-2979
E-mail: disasters@delphi.com
Website: http://www.disasters.org/dera.html

DisasterCom
Quarterly periodical–$15/year (subscription part of membership fee)
Disaster Emergency Response Association
Address, fax, e-mail and website are the same as listed above.
Tel: (414) 587-3636 or (970) 532-3362

Disaster Recovery Journal ✓
Quarterly periodical–free to all qualified personnel involved in managing, preparing, or supervising contingency planning; otherwise $10/year U.S.; $24/year Canada and Mexico; $47/year elsewhere
11337 St. John's Church Road, St. Louis, MO 63123
Tel: (314) 894-0276
Fax: (314) 894-7474
E-mail: drj@drj.com
Website: http://www.drj.com

Disaster Resource Guide ✓
Annual periodical–free to qualified individuals
P.O. Box 153243, santa Ana, CA 92735
Tel: (800) 826-2201 or (714) 558-8940
Website: http://www.disaster-resource.com

Disaster Response Network ✓
Three times per year periodical–free
Kelly Kennai, American Psychological Association, Practice Directorate, 750 First Street, N.E., Washington, DC 20002-4242
Tel: (202) 336-5898
E-mail: practice@apa.org
Website: http://www.apa.org

Disasters: Preparedness and Mitigation in the Americas ✓
Quarterly–free (available in English and Spanish)
Pan American Health Organization, 525 23rd Street, N.W. Washington, DC 20037
Tel: (202) 974-3520
Fax: (202) 775-4578
E-mail: disaster@paho.org
Website: http://www.paho.org

Emergency Preparedness Canada
Chris Tucker, Director, Evaluation and Analysis
122 Bank Street, Second Floor, Jackson Building, Ottawa, Ontario, Canada K1A 0W6
Tel: (613) 991-7071
Fax: (613) 996-0995
E-mail: devall@fox.nstn.ca

Emergency Preparedness Digest
Quarterly periodical–$26 (Canadian)/year Canada; $26 (U.S.)/year in
U.S. and outside Canada
Canada Communications Group-Publishing, Ottawa, Canada K1A 0S9
Tel: (819) 956-4802
Fax: (819) 994-1498

Emergency Preparedness News
Biweekly periodical–$327/year, add $13 for airmail postage outside
the U.S.
BPI, 951 Pershing Drive, Silver Spring, MD 20910-4464
Tel: (301) 587-6300 or (800) 274-6737
Fax: (301) 587-1081
E-mail: bpihlth@bpinews.com
Website: http://www.bpinews.com

Federal Emergency Management Agency
James L. Witt, Director
500 C Street, S.W., Washington, DC 20472
Tel: (202) 646-3923
Fax: (202) 646-3930
E-mail: eipa@fema.gov
Website: http://www.fema.gov
Office of Emergency Information and Public Affairs
Vallee D. Bunting, Director; Phil Cogan, Deputy Director
Tel: (202) 646-4600
Fax: (202) 646-4086
(Same E-mail and Website as above)

George Washington University
Institute For Crisis And Disaster Management, Research And Education
John Harrald, Director
George Washington University, Virginia Campus, 20101
Academic Way, Room 220, Ashburn, VA 22011
Tel: (202) 994-7153
E-mail: harrald@seas.gwu.edu

International Association Of Emergency Managers
(Formerly National Coordinating Council On Emergency Management)
Elizabeth Armstrong, Executive Director
111 Park Place, Falls Church, VA 22046-4513
Tel: (703) 538-1795
Fax: (703) 241-5603
E-mail: iaem@aol.com

International Journal of Mass Emergencies and Disasters
Three times per year periodical–$48/year for institutions, $20/year for individuals
David Neal, Institute of Emergency Administration and Planning, P.O. Box 13438, University of North Texas, Denton, TX   76203

Local Authorities Confronting Disasters And Emergencies
Avi Rabinovich
The Office of the Secretary General, Union of Local Authorities in Israel, 3 Heftman Street, Tel Aviv 61200, Israel, P.O. B. 20040
Tel: 972-3-6955024 or 6919241
Fax: 972-3-6967447
E-mail: ulais@netvision.net.il

Macquarie University, Natural Hazards Research Centre
Russell Blong, Director
School of Earth Sciences, New South Wales 2109, Austrailia
Tel: 61-(0)2-9850-9683
Fax: 61-(0)2-9850-9394
E-mail: rblong@laurel.ocs.mq.edu.au
Website: http://www.es.mq.edu.au/NHRC

National Centre For Disaster Management
IIPA, Indraprastha Estate, Ring Road, New Delhi 11002, India
Fax: (+91-11) 331-9954

National Emergency Management Association
David Rodham
P.O. Box 11910, Lexington, KY   40578-1910
Tel: (606) 244-8000
Fax: (606) 244-8239
E-mail: thembree@csg.com
Website: http://www.nemaweb.org

National Emergency Training Center
16825 South Seton Avenue, Emmitsburg, MD   21727
Tel: (301) 447-1000

Natural Hazards: Journal of the International Society for the *Prevention and Mitigation of Natural Hazards*.
Bimonthly periodical–$214/year
Kluwer Academic Publishers, P.O. Box 322, 3300 AH Dordrecht, the Netherlands, or P.O. Box 958, Accord Station, Hingham, MA   02018-0358

National Oceanic And Atmospheric Administration
(See also listing under Severe Weather)
Central Library
Carol Watts, Chief
1315 East West Highway, Second Floor, Silver Spring, MD 20910
Tel: (301) 713-2600
Fax: (301) 713-4598
E-mail: reference@nodc.noaa.gov
Website: http://www.bfrl.nist.gov

National Research Institute For Earth Science And Disaster Prevention
Tsuneo Katayama
3-1, Tennodai, Tsukiba-shi, Ibaraki-ken, 305 Japan
Tel: (81-298) 51-1611
Fax: (81-298) 51-1622, (81-298) 51-3246
E-mail: plansec@mhintos.ess.bosai.go.jp

Natural Hazards Observer
Bimonthly periodical–free within the U.S., $15/year, elsewhere.
Publications Clerk, Natural Hazards Research and Applications Information Center, IBS #6, Campus Box 482, University of Colorado, Boulder, CO 80309-0482
Tel: (303) 492-6819
Fax: (303) 492-2151
E-mail: jclark@colorado.edu
Website: http://www.colorado.edu/hazards

Natural Hazards Society
S. Venkatesh, Membership Director
Atmospheric Environment Service, 4905 Dufferin Street, Downsview, Ontario, Canada M3H ST4
Tel: (416) 739-4911
Fax: (416) 739-4288
E-mail: vankatesh@am.dow.on.doe.ca

Natural Hazards Society Newsletter
Quarterly periodical–free with membership
Seana O'Brien, Editor, National Tidal Facility, P.O. Box 2100, Adelaide 5001, Australia
Tel: 61-8-8201-7614
Fax: 61-8-8201-7523
E-mail: seana@pacific.ntf.flinders.edu.au

United Nations
International Decade For Natural Disaster Reduction (IDNDR) Secretariat
Philipppe Boullé, Director; Natalie Domeisen, Promotion Officer
United Nations, Palais des Nations, CH-1211 Geneva 10, Switzerland
Tel: (41-22) 798 68 94
Fax: (41-22) 733 86 95
E-mail: Natalie.Domeisen@dha.unicc.org

University Of Colorado-Boulder
Natural Hazards Research and Applications Information Center
Campus Box 482, Boulder, CO 80309-0482
Tel: (303) 492-6818
Fax: (303) 492-2151
E-mail: hazctr@colorado.edu
Website: http://www.colorado.edu/hazards

University Of Delaware
Disaster Research Center
Newark, DE 19716
Joanne Nigg or Kathleen Tierney, Co-Directors
Susan Castelli, Librarian
Tel: (302) 831-6618
Fax: (302) 831-2091
E-mail: joanne.nigg@mvs.udel.edu or tierney@udel.edu or
Susan.Castelli@mvs.udel.edu
Website: http://www.udel.edu/DRC/homepage.htm

University Of South Carolina
Hazards Research Laboratory
Susan Cutter, Director
Department Of Geography, Columbia, SC 29208
Tel: (803) 777-1699
Fax: (803) 777-4972
E-mail: uschrl@ecotopia.geog.sc.edu
Website: http://www.cla.sc.edu/geog/hrl/home.html

University Of Wisconsin
Disaster Management Center
Don Schramm, Director
Department of Engineering Professional Development, 432 North
Lake Street, Madison, WI 53706
Tel: (608) 262-5441
Fax: (608) 263-3160
E-mail: dmc@engr.wisc.edu
Website: http://epdwww.engr.wisc.edu/dmc/

U.S. Coast Guard, National Response Center
Jeffrey Ogden
2100 Second Street, S.W., Room 2611, Washington, DC 20593
Tel: (202) 267-2185 or Hotline: (800) 424-8802
Fax: (202) 267-2165
E-mail: jogden@comdt.uscg.mil
Website: http://www.dot.gov/dotinfo/uscg/hq/nrc

Avalanches and Landslides

American Association Of Avalanche Professionals
Peggy Rickets, Executive Secretary
P.O. Drawer 2757, Truckee, CA 96160
Tel: (916) 587-9653
Fax: (916) 587-1727
E-mail: 7141351@mcimail.com

Avalanche Review
Monthly periodical between November and April. $20/year. (American Association Of Avalanche Professionals)
Don Bachman, Executive Director
P.O. Box 1032, Bozeman, MT 59771-1032
Tel: (406) 587-3830
Fax: (406) 586-4307
E-mail: avalpro@theglobal.net
Website: http://www.avalanche.org

International Landslide Research Group Newsletter
Irregularly – $2/three issues in U.S., Canada, Mexico, $4 elsewhere.
Earl E. Brabb
International Landslide Reasearch Group
3262 Ross Road, Palo Alto, CA 95030
Tel: (415) 329-5140

National Landslide Information Center.
Division of U.S. Geological Survey
Lynn M. Highland, Director
MS-966, P.O. Box 25046, Federal Center, Denver, CO, 80255-0046.
Tel: (800) 654-4966
Fax: (303) 279-8600
E-mail: nlic@usgs.gov
Website: http://gldage.cr.usgs.gov/html

Cyclones, Hurricanes and Typhoons
(See also: Severe Weather and Tornadoes)

Clemson University
Coastal Hazards Assessment and Mitigation Program
Dept. of Civil Engineering, Clemson, SC 29634-0911
Benjamin L. Sill, Director
Denise James, Executive Support Specialist
Tel: (803) 656-0488
E-mail: champ@eng.clemson.edu
Website: http://www.champ.eng.clemson.edu/

Duke University
Program For The Study Of Developed Shorelines
Orrin Pilkey, Director
341 Old Chemistry, Box 90228, Durham, NC 27706
Tel: (919) 684-4238
Fax: (919) 684-5847

Florida International University,
International Hurricane Center
Tom David or Walter Peacock, University Park Campus, Miami, FL 33199
Tel: (305) 348-1607
Fax: (305) 348-1605
E-mail: hurrican@fiu.edu
Website: http://www.fiu.edu/~hurrican/

National Weather Service
Tropical Prediction Center
National Hurricane Center, 11691 S.W. 17th Street, Miami, FL 33165-2149
Robert Burpee, Director, Tel: (305) 229-4402
Vivian Jorge, Administrative Officer, Tel: (305) 229-4403
E-mail: rburpee@nhc.noaa.gov
Website: http://www.nhc.noaa.gov

Storm Data
Monthly periodical – $53/year, plus $11/original order for service and handling.
National Climatic Data Center
151 Patton Avenue, Room 120, Asheville, NC 28801
Tel: (704) 271-4258
Fax: (704) 271-4876
E-mail: order@ncdc.noaa.gov
Website: http://www.ncdc.noaa.gov

Drought and Famine

Drought Network News. ✓
Triannual Newsletter – Free
University of Nebraska-Lincoln, 239 L.W. Chase Hall, P.O. Box 830749,
Lincoln, NE 68583-0728
Tel: (402) 472-6707
Fax: (402) 472-6614
E-mail: ndmc@enso.unl.edu

European Drought Mitigation Network
Institute of Hydrology
Jim Wallace
Wallingford, Oxfordshire OX10 8BB, U.K.
Tel: 44(0) 1491 838800
Fax: 44(0) 1491 692430
E-mail: jsw@unixa.nercwallingford.ac.uk
or
Tom Downing
Environmental Change Unit, University of Oxford, Oxford OX1 3TB, U.K.
Tel: 44(0) 1865 281180
Fax: 44(0) 1865 281181
E-mail: tom.downing@ecu.ox.ac.uk

United Nations
World Food Programme
Food Aid Information Group
Information System Service, Via Cristoforo Colombo 426, I-00145
Rome, Italy.
George-Andre Simon, Chief
Tel: (396) 52282796
Fax: (396) 52282451
E-mail: simong@unicc.bitnet or simong@wfp.org

University of Arizona
Office Of Arid Lands Studies And Arid Lands Information Center
Barbara Hutchinson, Director and Librarian
1955 East 6th Street, Tucson, AZ 85719-5224.
Tel: (520) 621-8578
Fax: (520) 621-3816
E-mail: barbarah@ag.arizona.edu
Website: http://ag.arizona.edu/OALS/oals/oals.html

University of Nebraska-Lincoln
International Drought Information Center
Department of Agricultural Meteorology
Donald A. Wilhite, Director
241 L.W. Chase Hall, Lincoln, NE 68583-0749
Tel: (402) 472-6707 or (402) 472-4270
Fax: (402) 472-6614
E-mail: ncmc@enso.unl.edu or dwilhite@enso.unl.edu
Website: http://enso.unl.edu/ndmc

University of Nebraska-Lincoln
National Drought Mitigation Center
Department of Agricultural Meteorology
Donald A. Wilhite, Director
239 L.W. Chase Hall, Lincoln, NE 68583-0749
(same telephone, fax, e-mail and website as above)

Earthquakes and Tsunami

Abstract Journal in Earthquake Engineering (AJEE)
Biannual Journal – $100, $125 outside North America
National Information Service for Earthquake Engineering
Earthquake Research Center
University of California-Berkeley, 1301 South 46th Street, Richmond,
CA 94804-4698.
Tel: (510) 231-3468
Fax: (510) 231-9461
E-mail: shirley@eerc.berkeley.edu
Website: http://www.eerc.berkeley.edu

California Seismic Safety Commission
Richard McCarthy, Executive Director
1900 K Street, Suite 100
Sacramento, CA 95814
Tel: (916) 322-4917
Fax: (916) 322-9476
Website: http://earthview.sdsu.edu/SSC/index.html

Cascadia Region Earthquake Work Group
University of Washington Seismology Laboratory
Pacific Northwest Seismic Network
Box 251650, Seattle, WA 98195-1650
Website: http://www.geophys.washington.edu/CREW/index.html

Central United States Earthquake Consortium
Tom Durham, Executive Director
2630 East Holmes Road, Memphis, TN 38118-8001
Tel: (901) 544-3570
Fax: (901) 544-0544
E-mail: wbalda@simeon.org
Website: http://www.simeon.org/msm.html

Charleston Southern University
Earthquake Education Center
Joyce Bagwell, Director
P.O. Box 118087, Charleston, SC 29423-8087
Tel: (803) 863-8088
Fax: (803) 863-7533
E-mail: jbagwell@awod.com

Earthquake Engineering Research Institute
Susan Tubbesing, Executive Director
499 14th Street, Suite 320 Oakland, CA 94612-1934
Tel: (510) 451-0905·
Fax: (510) 451-5411
E-mail: skt@eeri,org
Website: http://gandalf.ceri.memphis.edu/~CUSEC/index.html

Earthquake Spectra
Quarterly periodical – $75/year for individuals, $120/year for institutions.
Earthquake Engineering Research Institute
499 14th Street, Suite 320, Oakland, CA 94612-1934
Tel: (510) 451-0905
Fax: (510) 451-5411
E-mail: eeri@eeri.org
Website: http://www.eeri.org

EERC News ✓
Quarterly Newsletter – Free
Earthquake Engineering Research Center (EERC)
University of California-Berkeley, 1301 South 46th Street, Richmond,
CA 94804
Tel: (510) 231-9554
Fax: (510) 231-9461
E-mail: janine@eerc.berkeley.edu
Website: http://www.eerc.berkeley.edu

EERI Newsletter
Monthly Newsletter – $115/yr (includes membership)
Earthquake Engineering Research Institute
499 14th Street, Suite 320, Oakland, CA 94612-1934
Tel: (510) 451-0905
Fax: (510) 451-5411
E-mail: eeri@eeri.org
Website: http://www.eeri.org

epiCenter News ✓
Quarterly Newsletter – Free
Richard Cook, U.S. Army Corps of Engineers
Earthquake Preparedness Center of Expertise, Room 1024, 333 Market
Street, San Fransisco, CA 94105-2195
Tel: (415) 977-8326
Fax: (415) 977-8346
E-mail: rcook@smtp.spd.usace.army.mil
Website: http://www.spn.usace.army.mil/earth.html

Fault Line Forum ✓
Quarterly Periodical – Free
Utah Geological Survey, P.O. Box 146100, Salt Lake City, UT 84114-6100.
Tel: (801) 537-3383
Fax: (801) 537-3400
E-mail: nrugs.bmayes@state.ut.us
Website: http://www.ugs.state.ut.us

International Tsunami Information Center
Charles S. McCreery, Director
Biannual periodical: "Tsunami Newsletter" ✓
Free to scientists, engineers, educators, community protection agencies,
and governments worldwide.
737 Bishop Street, Suite 2200, Honolulu, HI 96813-3213
Tel: (808) 532-6422
Fax: (808) 532-5576
E-mail: itic@itic.noaa.gov

National Center For Earthquake Engineering Research
George C. Lee, Director
State University of New York at Buffalo
Red Jacket Quadrangle, Box 610025, Buffalo, NY 14261-0025
Tel: (716) 645-3391
Fax: (716) 645-3399
E-mail: nernceer@ubvms.cc.buffalo.edu
Website: http://nceer.eng.buffalo.edu

Seismological Society Of America
Susan Newman, Executive Director
201 Plaza Professional Building, El Cerrito, CA 94530-4003
Tel: (510) 525-5474
Fax: (510) 545-7204
E-mail: info@seismosoc.org
Website: http://www.seismosoc.org

Southeast Missouri State University
Center For Earthquake Studies
Chris Sanders, One University Plaza, Cape Girardeau, MO 63701-4799
Tel: (314) 654-2019
Fax: (314) 654-2316
E-mail: csanders@rose.gs.st.semo.edu

Southern California Earthquake Center
Quarterly Newsletter – $25/year
Southern California Earthqake Center
University of Southern California
University Park, Los Angeles, CA 90089-0742
Tel: (213) 740-1560
Fax: (213) 740-0011
E-mail: scecinfo@usc.edu
Website: http://www.usc.edu/gp/scec

Stanford University
John A. Blume Earthquake Engineering Center
Anne S. Kiremidjian
Department of Civil Engineering, Building 540,
Stanford University, Stanford, CA 94305-4020
Tel: (415) 723-4150
Fax: (415) 723-9755
E-mail: ask@ce.stanford.edu

Tsunami Society
Biannual periodical: "Science of Tsunami Hazards"
$20/year in the U.S. or free with membership
P.O. Box 25218, Honolulu, HI 96825
Website: http://www.ccalmr.ogi.edu/STH/society.html

United States Geological Survey
National Earthquake Information Center
Waverly Person, Director
MS-967, P.O. Box 25046, Federal Center, Denver, CO 80225.
Tel: (303) 273-8500 (Operations)

Tel: (303) 273-8516 (Earthquake Information)
Fax: (303) 273-8450
E-mail: sedas@gldfs.cr.usgs.gov
Website: http://wwwneic.cr.usgs.gov/

University Of California-Berkeley
Earthquake Engineering Research Center and National Information
Service for Earthquake Engineering
Chuck James, University Of California, 1301 South 46th Street, Richmond,
CA 94804
Tel: (510) 231-9401
Fax: (510) 231-9461
E-mail: cjames@eerc.berkeley.edu
Website: http://nisee.ce.berkeley.edu

University Of Memphis
Center For Earthquake Research And Information, Seismic Resource Center
Jill Stevens-Johnston, Director
Campus Box 526590, Memphis, TN 38152
Tel: (901) 678-2007
Fax: (901) 678-4734
E-mail: stevens@ceri.memphis.edu
Website: http://www.ceri.memphis.edu/

University Of Southern California
Southern California Earthquake Center
Jill Andrews
Department of Earth Sciences
University Park Los Angeles, CA 90089-0742
Tel: (213) 740-3459
Fax: (213) 740-0011
E-mail: jandrews@usc.edu
Website: http://www.usc.edu/go/scec

Western States Seismic Policy Council
Steven Ganz, Executive Director
121 Second Street, 4th Floor, San Fransisco, CA 94105
Tel: (415) 974-6435
Fax: (415) 974-1747
E-mail: wsspc@wsspc.org
Website: http://www.wsspc.org

Floods

Association Of State Dam Safety Officials
Lori Spragens, Executive Director
450 Old East Vine, Second Floor, Lexington, KY 40507
Tel: (606) 257-5140
Fax: (606) 323-1958
E-mail: 72130.2130@compuserve.com
Website: http://ourworld.compuserve.com/homepages/ASDSO/

Association Of State Floodplain Managers
Larry Larson, Execturive Director
4299 West Beltline Highway, Madison, WI 53711
Tel: (608) 274-0123
Fax: (608) 274-0696
E-mail: asfpm@execpc.com

Middlesex University
Flood Hazard Research Centre
Dennis Parker
Queensway, Enfield, Middlesex EN3 4SF, U.K.
Tel: +44 181 362 5359
Fax: +44 181 362 5403
E-Mail: fhrc1@mdx.ac.uk
Website: http://www.mdx.ac.uk/www/gem/fhrc.htm

National Association Of Flood And Stormwater Management Agencies
Susan Gilson, Executive Director
1225 Eye Street, N.W., Suite 300, Washington, DC 20005
Tel: (202) 682-3761 ext. 239
Fax: (202) 842-0621

U.S. Army Corps Of Engineers
Flood Plain Management Services And Coastal Resources Branch
20 Massachusetts Avenue, N.W., Washington, DC 20314
Tel: (202) 272-0169
Fax: (202) 272-1972
Website: http://www.hq.usace.army.mil/cecw/planning/main.htm

U.S. Geological Survey
National Water Information Center
427 National Center, Reston, VA 20192
Tel: (800) 426-9000
E-mail: h2oinfo@usgs.gov
Website: gtto://h2o.usgs.gov

U.S. Water News
Monthly periodical, $54/year, U.S., $64/year, Canada, $99/year elsewhere.
Circulation Department, 230 Main Street, Halstead, KS 67056
Tel: (316) 835-2222
Fax: (316) 835-2223
E-mail: uswatrnews@aol.com
Website: http://www.dcn.davis.ca.us/~uswn/homepage.html

Watermark: The NFIP Newsletter ✓
Biannual periodical – Free
Federal Emergency Management Agency
National Flood Insurance Program, Public Affairs Office
10115 Senate Drive, Lanham, MD 20706
Tel: (301) 731-5300

Severe Weather

(See also: Cyclones and Tornadoes)

American Association Of Wind Engineers
Ahsan Kareem
Department of Civil Engineering and Geological Sciences
University of Notre Dame, Notre Dame, IN 46556-0767
Tel: (219) 631-6648 or 631-7385
Fax: (219) 631-8007
E-Mail: kareem@navier.ce.nd.edu

American Meteorological Society
Richard Hallgren, Executive Director
45 Beacon Street, Boston, MA 02108
Tel: (617) 227-2425
Fax: (617) 742-8718
E-mail: hallgren@ametsoc.org
Website: http://www.ametsoc.org/AMS

American Weather Observer
Monthly periodical – $24.95/year
401 Whitney Boulevard, Belvidere, IL 61008-3772
Tel: (815) 544-9811
Fax: (815) 544-6334
E-mail: awowx@aol.com
Website: http//members.aol.com/larrypahl/awo.htm

Aware Report
Quarterly Periodical – Free
Linda Kremkau, National Weather Service
1325 East West Highway, Room 14370, Silver Spring, MD 20910
Tel: (301) 713-0090, ext. 118
E-mail: linda.kremkau@noaa.gov
Website: http://www.nws.noaa.gov/om/aware.pdf

Colorado State University
Fluid Mechanics and Wind Engineering Program, Fluid Dynamics and
Diffusion Laboratory
Robert N. Meroney, Director
Department of Civil Engineering, Fort Collings, CO 80523
Tel: (970) 491-8574
Fax: (970) 491-8671
E-mail: meroney@engr.colostate.edu
Website: http://www.lance.colostate.edu/depts/ce/netscap/depts/
fluid_mechanics

National Lightning Safety Institute
Richard Kithil, Executive Director
891 North Hoover Avenue, Louisville, CO 80027
Tel: (303) 666-8817
Fax: (303) 666-8786
E-mail: rich@lightningsafety.com
Website: http://www.lightningsafety.com

National Oceanic And Atmospheric Administration
National Climatic Data Center
Primary public contact point and climatic data ordering service
151 Patton Avenue, Asheville, NC 28801
Tel: (704) 271-4682
Fax: (704) 271-4876
E-mail: ncdc@noaa.gov
Website: http://www.ncdc.noaa.gov

National Oceanic And Atmospheric Administration
National Severe Storms Laboratory
Douglas Forsyte, Acting Director
1313 Halley Circle, Norman, OK 73069
Tel: (405) 366-0427
Fax: (405) 366-0472
E-mail: forsyte@nssl.uoknor.edu
Website: http://www.nssl.uoknor.edu

National Weather Service
Industrial Meteorology Staff (W/IM)
Silver Spring, Metro Center 2, Station 18462, 1325 East West Highway,
Silver Spring, MD 20910
Tel: (301) 713-0258
Fax: (301) 713-0610
Website: http://www.nws.noaa.gov/im/index.html

National Weather Service
National Centers For Environmental Prediction, Cimate Prediction Center
Richard Tinker, Editor, Weekly Climate Bulletin
W/NMC53, Room 805, World Weather Building, Washington, DC 20223
Tel: (301) 763-4670
Fax: (301) 763-8125
E-mail: tinker@climon.wwb.noaa.gov
Website: http://nic.sb4.noaa.gov

National Weather Service
Aviation Weather Center
David R. Rodenhuis, Director
Federal Building, Room 1728, 601 East 12th Street, Kansas City, MO 64106
Tel: (816) 426-5922
Fax: (816) 426-3453
E-mail: david.rodenhuis@noaa.gov

National Weather Service
Office Of Meteorology, Warnings and Forecast Branch
Donald R. Wernly
W/OM11, Room 14414, 1325 East West Highway, Silver Spring, MD 20910
Tel: (301) 713-0090
Fax: (301) 713-7598
E-mail: don.wernly@noaa.gov

Texas Tech University,
Department Of Civil Engineering
Wind Engineering Research Center
Kishor C. Mehta, Director
April MacDowell, Research Coordinator
Box 41023, Lubbock, TX 79409-1023
Tel: (806) 742-3476
Fax: (806) 742-3446
E-Mail: amacdowell@coe2.coe.ttu.edu

United Nations
World Meteorological Organization (WMO)
Haleh Kootval, World Weather Watch
P.O. Box 2300, CH-1211 Geneva 2, Switzerland
Tel: (41-22) 730 81 11, ext 221
Fax: (41-22) 734 23 26
Telex: 41 41 99 OMM CH
E-mail: ipa@www.wmo.ch
Website: http://www.wmo.ch

Universtiy Of East Anglia, Climatic Research Unit
Trevor Davies, Director
Norwich NR4 7TJ, U.K.
Tel: +44-1603-592722
Fax: +44-1603-507784
E-mail: t.d.davies@uea.ac.uk
Website: http://www.cru.uea.ac.uk

University Of Oxford, Environmental Change Unit
Thomas E. Downing
Climate Impacts and Responses, University of Oxford, Oxford OX1 3TB, U.K.
Tel: +44 1865 281180
Fax: +44 1865 2811811
E-Mail: Tom.Downing@ecu.ox.ac.uk
Website: http://info.ox.ac.uk/departments/ecu

U.S. Army Corps Of Engineers
Cold Regions Research and Engineering Laboratory
Nancy Liston, Librarian
72 Lyme Road, Hanover, NH 03755-1290
Tel: (603) 646-4221
Fax: (603) 646-4712
E-mail: nliston@crrel.usace.army.mil
Website: http://www.crrel.usace.army.mil

Weatherwise
Bimonthly periodical – $54/year for institutions; $32/year for individuals; $12/year additional for subscriptions outside the U.S.
Heldref Publications
1319 18th Street, N.S., Washington, DC 20036-1802
Tel: (800) 356-9753

The Wind Engineer
Quarterly periodical – $25/year (included in membership fee)
American Association For Wind Engineering
P.O. Box 1159, Notre Dame, IN 46556-1159

Tel: (219) 631-6648
Fax: (219) 631-9236
E-mail: kareem@navier.ce.nd.edu

Tornadoes
(See also: Cyclones and Severe Weather)

The Tornado Project
Tom Grazulis, Director
P.O. Box 302, St. Johnsbury, VT 03819
E-mail: tornproj@plainfield.bypass.com
Website: http://www.tornadoproject.com/

Volcanoes

Bulletin Of The Global Volcanism Network
Monthly periodical – $20/year in U.S., $32 elsewhere
American Geophysical Union, 2000 Florida Avenue N.W., Washington, DC 20009
Tel: (202) 468-6900 or (800) 966-2481
Fax: (202) 328-0566
E-mail: orders@kosmos.agu.org
Website: http://www.agu.org

Global Volcanism Program
Jim Luhr
National Museum of Natural History
MRC 129, Smithsonian Institution, Washington, DC   20560
Tel: (202) 357-4809
Fax: (202) 357-2476
E-mail: gvp@volcano.si.edu
Website: http://www.volcano.si.edu/gvp/

International Association Of Volcanology And Chemistry Of The Earth's Interior
IAVCEI Secretariat, P.O. Box 185, Campbell ACT 2612, Australia
Fax: 61 6 249-9986
E-mail: cgidding@agso.gov.au

U.S. Geological Survey
Volcano Hazards Program
Marianne Guffanti
MS-905, National Center, Reston, VA   20192
Tel: (703) 648-6708
Fax: (703) 648-6717
E-mail: guffanti@usgs.gov
Website: http://vulcan.wr.usgs.gov/Vhp/framework.html

Wildland Fires

Current Titles In Wildland Fire
Monthly periodical – $50/year
International Association Of Wildland Fire
(see organization listing for address and other information)

Fire International
Bimonthly periodical – £61.90/year in U.K., £71.80/year overseas,
£114.90/year in U.S.
Argus Business Media Ltd., Queensway House, 2 Wueensway, Redhill,
Surrey RH1 1QS, U.K.
Tel: +44(0) 1737 768611
Fax: +44(0) 1737 761685
E-mail: fireint@envsserv.demon.co.uk

IAFC On Scene
Twice monthly periodical – free with membership, or $60/year in U.S.,
$70/year elsewhere
International Association Of Fire Chiefs, attn: Tim Elliott
(see organization for address, fax, and website)
Tel: (703) 273-0911, ext. 307
E-mail: onscene@aol.com

International Association Of Fire Chiefs
Michael O. Forgy
4025 Fair Ridge Drive, Fairfax, VA   22033-2868
Tel: (703) 273-0911
Fax: (703) 273-9363
E-mail: iems@connectinc.com
Website: http://www.ichiefs.org

International Association Of Wildland Fire
Jason Greenlee, Executive Director
P.O. Box 328, Fairfield, WA   99012

Tel: (509) 283-2397
Fax: (509) 283-2264
E-mail: greenlee@cet.com
Website: http://www.neotecinc.com/wildfire

International Journal Of Wildland Fire
Quarterly periodical – $105/year
International Association of Wildland Fire
(see organization listing for address and other information)

National Emergency Training Center
16825 South Seton Avenue, Emmitsburg, MD 21727
Tel: (301) 447-1000
National Fire Academy
Denis G. Oniel, Superintendent (301) 447-1117
James F. Coyle, Deputy Superintendent (301) 447-1118
             and
U.S. Fire Administration
Carrye B. Brown, Administrator
Tel: (301) 447-1018
Fax: (301) 447-1270

U.S. Department Of Agriculture
Forest Service, Fire, and Aviation Management
Denny Truesdale, Emergency Disaster Coordinator
P.O. Box 96090, Washington, DC 20090-6090
Tel: (202) 205-1485
Fax: (202) 205-1272

Wildfire
Quarterly periodical – $30/year
International Association of Wildland Fire
(see organization listing for address and other information)

Wildfire News and Notes
Four to six times a year periodical — Free
National Fire Protection Association, 1 Batterymarch Park, P.O. Box
9101, Quincy, MA 02269-9101
Tel: (617) 770-3000
Website: http://www.firewise.org/pubs/wnn

Bibliography

Allaby, Michael. *Dangerous Weather: Blizzards.* New York: Facts On File, 1997.

Allaby, Michael. *Dangerous Weather: Tornadoes.* New York: Facts On File, 1997.

Barrett, Norman. *Picture Library: Volcanoes.* New York: Franklin Watts, 1989.

Bolt, Bruce A. *Earthquakes And Geological Discovery.* New York: Scientific American Library, 1993.

Cable, Mary A. *The Blizzard Of '88.* New York: Atheneum, 1988.

Cornell, James. *The Great International Disaster Book.* New York: Charles Scribner's Sons, 1976.

Dolan, Edward F. *Drought The Past, Present, And Future Enemy.* New York: Franklin Watts, 1990.

Erickson, Jon. *Violent Storms.* Pennsylvania: Tab Books, Inc., 1988.

Fisher, David E., *The Scariest Place On Earth: Eye To Eye With Hurricanes.* Random House, New York, 1994.

Fodor, R.V. *Earth Afire! Volcanoes And Their Activity.* New York: William Morrow And Company, 1981.

Fradin, Dennis Brindell. *Disaster! Drought.* Chicago: Children's Press, 1983.

Fradin, Dennis Brindell. *Disaster! Earthquakes.* Chicago: Children's Press, 1982

Fradin, Dennis Brindell. *Disaster! Volcanoes.* Chicago: Children's Press, 1982.

Frasier, Colin. *The Avalanche Enigma.* Chicago: Rand McNally and Company, 1966.

Frazier, Kendrick. *The Violent Face Of Nature.* New York: William Morrow & Company, Inc., 1979.

Kals, W.S. *Riddle Of The Winds.* New York: Doubleday & Company, Inc., 1977.

Laskin, David. *Braving The Elements.* New York: Doubleday, 1996.

Lauber, Patricia. *Hurricanes: Earths Mightiest Storms.* Scholastic Press, New York, 1996.

Levy, Matthys and Salvadori, Mario. *Why The Earth Quakes.* New York: W.W. Norton & Company, 1995.

McCullough, David G. *The Johnstown Flood.* New York: Simon And Schuster, 1968.

Micallef, Mary. *Floods and Droughts.* Illinois: Good Apple, 1985.

National Geographic Society, Book Division. *Restless Earth: Disasters of Nature.* Washington DC: National Geographic Society, 1997.

National Geographic Society, Special Publications Division. *Powers Of Nature.* Washington DC: National Geographic Society, 1978.

Office Of Emergency Preparedness. *Disaster Preparedness. Report To Congress.* Washington, D.C.: Government Printing Office, 1972.

Peissel, Michael and Allen, Missy. *The Encyclopedia of Danger - Dangerous Natural Phenomena.* Chelsea House Publishers, New York, 1993.

Posey, Carl A. *The Living Earth Book of Wind And Weather.* New York: The Reader's Digest Association, Inc. 1994.

Ritchie, David. *Superquake!* New York: Crown Publishers, Inc., 1988.

Robinson, Andrew. *Earth Shock: Hurricanes, Volcanoes, Earthquakes, Tornadoes, And Other Forces Of Nature.* New York: Thames And Hudson Ltd., 1993.

Rossbacher, Lisa A. *Recent Revolutions In Geology.* New York: Franklin Watts, 1986.

Thomas, Gordon and Witts, Max Morgan. *The San Francisco Earthquake.* New York: Stein and Day, 1971.

Uman, Martin A. *All About Lightning.* New York: Dover Publications, Inc., 1971.

Vogt, Gregory. *Predicting Volcanic Eruptions.* New York: Franklin Watts, 1989.

Walker, Jane. *Avalanches And Landslides.* New York: Gloucester Press, 1992.

Watson, Benjamin A. *The Old Farmer's Almanac Book Of Weather and Natural Disasters.* New York: Random House, 1993.

Index

S

To order additional copies of this book, send requests and payment to:

MedCap
Orders/Distribution
P.O. Box 2085
Clifton, NJ 07055

Price for single copies is $12.95 plus $3.00 shipping. Discounts available for larger orders. For more information, write to the address above or send an e-mail to: medcap.pub@juno.com

*New Jersey residents must include 6% sales tax.